Abiotic and Biotic Stress Management in Plants

About the Editors

Dr. Bhav Kumar Sinha was born on 12th September 1975 at Haiderchak, Nalanda, Bihar. He did his Master's Degree in Plant Physiology from Banaras Hindu University, Varanasi in 1999 and Ph.D. (Plant Physiology) from Chaudhary Charan Singh Haryana Agricultural University, Hisar in 2004. In July 2008, he joined Sher-e-Kashmir University of Agricultural Sciences and Technology of Jammu, Jammu & Kashmir as Asst. Professor cum-Junior Scientist. Since then he is serving this University as Plant Physiologist and has given a direction in teaching, Plant Physiological research and extension. Dr. Sinha's research areas are Stress Physiology and Hormonal Physiology. He has published more than 23 original research paper in Indian and international journals, seven book chapters in different book and two practical manual. He is life member of serveral professional society and has handled one externally final project.

Reena (D.O.B – 01/12/1975) working as Senior Scientist (Entomology) at ACRA, Dhiansar, SKUAST-J, has done her B.Sc. (Agriculture) from Banaras Hindu University, Varanasi, M.Sc. (Agril. Entomology) from University of Agricultural Sciences, Dharwad and secured Ph.D. (Entomology) degree from C.C.S. Haryana Agricultural University, Hisar. She has qualified NET conducted by A.S.R.B., New Delhi and CSIR, New Delhi. She is also the recipient of Junior Research Fellowship (ICAR) during MSc. (Ag) degree program and Department of Science and Technology (DST), Government of India, Young Scientist Project Award (2009- 2012) under fast track scheme for young scientists. She is life member of several professional societies and has handled two externally funded project as PI and two as Co-PI. She has delivered several expert lectures and has published 25 research papers in journals of national and international repute.

Dr. Surendra Prasad working as Jr. Scientist (SMS) Entomology, Krishi Vigyan Kendra, Manjhi, Saran, Dr. Rajendra Prasad Central Agricultural University, Pusa, Bihar - 841313 did his B. Sc. (Ag) and M. Sc. (Ag) in Agricultural Entomology from C. S. Azad University of Agriculture and Technology, Kanpur and Ph. D. degree in Entomology and Agricultural Zoology from Institute of Agricultural Sciences, Banaras Hindu University, Varanasi. Dr. Prasad has over six years experience as Research

(Research Associate) in “Insect Biosystematics” and eight years of experience as a plant protection specialist in KVK. He has published more than eighteen research papers, ten book chapters and more than twenty popular extension folders. More than 10 research abstracts and full paper presentation in national and international seminar/symposia to his credit. More than fifteen radio and TV talk have been delivered by him.

Abiotic and Biotic Stress Management in Plants

Volume-II: Biotic Stress

Bhav Kumar Sinha
Reena
Surendra Prasad

NEW INDIA PUBLISHING AGENCY
New Delhi – 110 034

First published 2022
by CRC Press
4 Park Square, Milton Park, Abingdon, Oxon, OX14 4RN

and by CRC Press
6000 Broken Sound Parkway NW, Suite 300, Boca Raton, FL 33487-2742

CRC Press is an imprint of Informa UK Limited

British Library Cataloguing-in-Publication Data
A catalogue record for this book is available from the British Library

Library of Congress Cataloging-in-Publication Data
A catalog record has been requested

ISBN: 978-1-032-26022-8 (hbk)
ISBN: 978-1-003-28613-4 (ebk)

DOI: 10.1201/9781003286134

MIX
Paper | Supporting responsible forestry
FSC™ C013985
FSC www.fsc.org

Printed in the United Kingdom
by Henry Ling Limited

Sher-e-Kashmir
University of Agricultural Sciences
and Technology - Jammu

Dr Jag Paul Sharma
Director Research

Foreword

Plants are often exposed to various abiotic and biotic stresses. They have developed specific mechanisms to adapt, survive and reproduce under these stresses. Together, these stresses constitute the primary cause of crop losses worldwide, reducing average yields of most major crop plants. Current climate change scenarios predict an increase in mean temperatures and drought that will drastically affect global agriculture in the near future. In agriculture abiotic and biotic stress not only cause huge reduction in crop yields but also increase cost of cultivation, reduce input use efficiency, impair quality of produce. A complete understanding on physiological and molecular mechanisms especially signaling cascades in response to abiotic and biotic stresses in tolerant plants will help to manipulate susceptible crop plants and increase agricultural productivity in the near future. The biology of plant cell is more complicated with any foreign stimulus from the environment; multiple pathways of cellular signaling and their interactions are activated. These interactions mainly evolved as mechanism to enable the plant systems to respond to stress with minimum and appropriate physio- biochemical processes. Advanced agricultural approaches are also required for sustainable solutions to the huge global problem of 'hidden hunger' and it may performed by the bio-fortification for increased micronutrient intakes and improved micronutrient status in the food. Management of cultural practices in conjunction with use of plant bio-regulators and chemicals will give a long way in management of various abiotic and biotic stresses.

Detailed discussion regarding cultural and agronomical management during stress condition will help not only to the researchers but also to the field functionaries and farmers. In this regard, an attempt has been made by Dr. Bhav Kumar Sinha and his team, Division of Plant Physiology, Faculty of Basic Sciences, Chatha, SKUAST-Jammu to come up with a volume entitled "Abiotic and Biotic Stress Management in Plants, Volume- II: Biotic Stress" containing about 12 Chapters.

This publication shall be of immense use to researchers, undergraduate and post graduate students along with other stake holders who is dealing with green ecosystem.

I congratulate the authors for their painstaking efforts in bringing out this publication.

Dr. Jag Paul Sharma

Main Campus Chatha, Jammu-180 009, J&K, INDIA

Tel: 0191-2263973
Mob: 09419134737
e-mail: jpsdr2015@gmail.com

Preface

Plants encounter a wide range of environmental insults during a typical life cycle and have evolved mechanisms by which to increase their tolerance of these through both physical adaptations, biochemical changes molecular and cellular changes that begin after the onset of stress. Environmental rudeness faces by the plants in the form of abiotic and biotic stress that seriously reduces their production and productivity. Approximately 70% of crops could have been lost due to both abiotic and biotic factors. Variety of distinct abiotic stresses, such as availability of water (drought, flooding), extreme temperature (chilling, freezing, heat), salinity, heavy metals (ion toxicity), photon irradiance (UV-B), nutrients availability, and soil structure are the most important features of and has a huge impact on growth and development and it is responsible for severe losses in the field and the biotic stress is an additional challenge inducing a negative pressure on plants and adding to the damage through herbivore attack or pathogen. Multiple stress exposure gives a possible outcome that Plant system develops tolerance to one environmental stress may affects the tolerance to another stress, for example, after exposure of plants to abiotic stress leading to enhanced biotic stress tolerance, wounding increases salt tolerance in tomato plants. In tomato plants, localized infection by *Pseudomonas syringae* pv. tomato (*Pst*) induces systemic resistance to the herbivore insect *Helicoverpa zea*.

Therefore, the subject of *Abiotic and Biotic Stress Management* is gaining considerable significance in the contemporary world. This book "Abiotic and Biotic Stress Management in Plants, Volume- II: Biotic Stress" deals with an array of topics in the broad area of biotic stress responses in plants focusing "*problems and their management*" by selecting some of the widely investigated themes. Chapter 1:Major insect-pest of cereal crops in India and their management, Chapter 2:Biotic stresses of major pulse crops and their management strategies, Chapter 3:Insect pest of oilseed crops and their management, Chapter 4:Biotic stresses of vegetable crops & their management, Chapter 5: Insect pests infesting major vegetable crops and their management strategies – I, Chapter 6: Insect pests infesting major vegetable crops and their management strategies – II, Chapter 7: Insect pests infesting major vegetable crops and their management strategies – III, Chapter 8: Fruit crops insect pests and their biointensive integrated pest management techniques, Chapter 9: Mass trapping of fruit flies using Methyl in Eugenol based traps, Chapter 10: Organic means of combating biotic stresses in plants, Chapter 11: Nematode problem in pulses and their management, Chapter 12: Recent Approaches in pest management of stored grain pests.

We fervently believe that this book will provide good information and understanding of biotic stress problems and their management in plants.

I would like to extend my gratitude to all contributors for their authoritative and up to date scientific information organized in a befitting manner. We thank the supporting staff of Division of Plant Physiology who have helped us in coming up with publication. The cooperation extended by Dr. J.P. Sharma, Director Research of the University is duly acknowledged. Valuable cooperation extended by Dr. S. A. Mallick, Dean, Faculty of Basic Sciences, in multifarious ways is gratefully acknowledged.

Last but not the least, I owe thanks to my son Krishna Sinha and Tanmay Sinha for taking care of me during this project.

Bhav Kumar Sinha
Reena
Surendra Prasad

Contents

List of Contributors

Ankit Kumar
Department of Entomology, CCS Haryana Agricultural University, Hisar- 125004

Arvind Singh Tetarwal
Subject Matter Specialist, CAZRI-KVK, Kukma Bhuj- 370105 Kachchh

Amandeep Kaur
DES (Ento), Punjab Agricultural University – Farm Advisory Service Scheme, Patiala – 141001

B. L. Jakhar
Centre of Excellence for Research on Pulses, S. D. Agricultural University
Sardarkrushinagar – 385 506

Kavita Bajaj
Department of Entomology, PAU, Ludhiana, 141004

Manoj Kumar Jat
Department of Entomology, CCS Haryana Agricultural University, Hisar - 125004

Reena
ACRA, SKUAST-Jammu, Dhiansar, Bari Brahmana-181133

R. M. Srivastava
College of Agriculture, G. B. Pant University of Agriculture and Technology, Pantnagar U.S. Nagar, Uttarakhand

Surendra Prasad
Krishi Vigyan Kendra, Manjhi, RAU, Saran, Bihar-841313

Surender Singh Yadav
Department of Entomology, CCS Haryana Agricultural University, Hisar- 125004

Smriti
DES (Ento), Punjab Agricultural University – Farm Advisory Service Scheme, Patiala – 141001

S. K. Maurya
Department of Vegetable Science, College of Agriculture, G. B. Pant University of Agriculture and Technology, Pantnagar, U.S. Nagar, Uttarakhand

Sandeep Singh
Department of Fruit Science, PAU, Ludhiana-141004

Tanuja Phartiyal
College of Agriculture, G.B. Pant University of Agriculture and Technology, Pantnagar U.S. Nagar, Uttarakhand

Virendra Kumar Singh
Department of Plant Pathology,College of Agriculture, Banda University of Agriculture and Technology-Banda-210001

1

Major Insect-Pest of Cereal Crops in India and Their Management

Surendra Prasad and Reena

Cereal crops are interchangeably called grain crops. In many publications and correspondence, they are simply called *grains* or *cereals*. As of 2012, the top 5 cereals in the world ranked on the basis of production tonnage are rice (paddy), wheat, maize (corn), barley and sorghum. These crops are also among the top 50 agricultural commodities in the world with maize ranking second next to sugarcane. Rice (paddy) ranks third, wheat - 4th, barley - 12th, and sorghum - 30th. Another cereal, millet, ranks no. 42 (FAO Stat 2014, updated Aug. 18, 2014). According to Chapman and Carter (1976), "a cereal is generally defined as a grass grown for its small, edible seed." They also explained that all cereals are angiosperms, monocots, and members of the grass family *Gramineae.* Similarly, Lantican (2001) defines cereal or grain crops as agronomic crops belonging to the grass family Gramineae or Poaceae which are utilized as staples; the word "cereal" is derived from the most important grain deity, the Roman Goddess Ceres. Among the food grains cereals occupies for most status in human food requirements. Cereals are grown under different agro-climatic (Upland, Low land & Rainfed) conditions and it is most important staple food of about 65% of Indian population. Cereals crop is attacked by a large number of insect pests right from initial to till the harvest of the crops. Brief descriptions of the major crops, major insect pest and appropriate control measures are given below:

Paddy (*Oryza sativa* L.)

Among the food grains paddy occupies fore-most status in human food requirements. Paddy is grown under different agro-climatic (Upland, Lowland &Rainfed) conditions and the crop is damaged by more than 100 sp of insect pests. This pest causes enormous grain yield losses, which may vary from 20-50% if not protected, India losses 30% yield in rice every year. It is most important staple food of about 65% of Indian population.

More than two dozens of insect pest have been recorded to infest rice crop. These pests do not appear regularly and cause varying degree of losses to rice in different stages in different parts of the country. Pest description and management of the pests are given below.

Common name	Scientific name	Family	Order
Yellow stem borer	*Scirpophaga incertulas* Walker	Pyraustidae	Lepidoptera
Rice Plant Hopper	*Nelaparvata lugens* Stal.	Delphacidae	Hemiptera
Green leaf hopper	*Nephotettix nigropictus* Stal.*N. virescens*	Cicadellidae	Hemiptera
Gundhi Bug	*Leptocorisa acuta* Fab.	Coreidae	Hemiptera
Rice Leaf folder	*Cnaphalocrocis medinalis* Guen.	Pyraustidae	Lepidoptera
Rice Hispa	*Dicladispa armigera* Oliv.	Hispidae	Coleoptera
Rice Grasshopper	*Hieroglyphus banian* Fab.	Acrididae	Orthoptera
Paddy Root Weevil	*Echinocnemus oryzae* Marshall	Curculuionidae	Coleoptera
Rice Caseworm	*Nymphula depunctalis* Guenee	Pyraustidae	Lepidoptera
Rice Gall Midge	*Orseolia oryzae* Wood-Mason	Cecidomyiidae	Diptera
Armyworm/ Climbing cutworm	*Mythimna separata* Walker*Agrotis ipsilon* (Hufn)	Noctuidae	Lepidoptera
Termites	*Odontotermies obesus Ramb.*	Termitidae	Isoptera
Plant lice (Aphid)	*Rhopalosiphum maidis Fitch.*	Aphididae	Hemiptera
Pyrilla	*Pyrilla perpusilla Wlk*	Lophopidae	Hemiptera
Rice skipper	*Pelopidas mathias Fb.*	Hesperiidae	Lepidoptera
Sorghum stem borer	*Chilo partellus (Sinhoe)*	Crambidae	Lepidoptera
Cob caterpillar	*Cryptoblabes angustipennella Hmpsn.*	Pyraustidae	Lepidoptera
White grub	*Holotrichia consanguinea Blanch*	Melolonthidae	Coleoptera
Rodent			

1. Yellow stem borer (YSB), *Scirpophaga incertulas* (Walker)

Host plant - Paddy is the main host plant for yellow stem borer.

Distribution - The insect is widely distributed in all Asian countries and is commonly known as yellow borer of rice, paddy stem borer or rice stem borer.

Nature of damage - This insect is monophagous. It is a serious pest in all the rice ecosystems, particularly in deep water rice in India. Freshly hatched larvae move downwards to leaf sheath and feed on inner tissue, with the advancement of growth and development larvae bore into stem bore into stem and feed on inner surface. Due to such feeding at vegetative stage the central leaf whorl unfold, turns brown and dries up which is termed as “Dead Heart”. Infestation after the panicle initiation result in drying of panicle which may not emerges at all and those that have already emerges do not produce grain and appears as “white head”.

Life cycle - Four development stages viz, egg – larva- pupa and adult are found in the life cycle. The female moth start egg laying at night near the tip on the upper surface of tender leaf in small masses and cover them with a felt-like buff coloured mass of hairs and scales. The number of eggs in each egg mass ranges from 15 - 80. Two to three clusters are laid by a female. It hatches in about 5-8 days. The first instar larva of yellow stem borer is about 1.5 mm long and pale white in colour with dark brown head and prothoracic shield. The newly hatched larvae move downward and wander about on the plant surfaces for 1 or 2 hours. They may hang down by silken thread, get blown off to other clumps or land on the water of paddy field, swim freely (as they have an air layer next to their skin) in search of seedlings and get to the plants. They enter the leaf sheath and feed upon the green tissues for 2-3 days, and then borer into the stem near the nodal region. They disperse from on plant to another and usually only one larva is found inside a stem. After a week it comes out, makes a case with leaf bits, drifts on water and attacks a fresh plant. There are about 6 instars and the full grown larva measures 20 mm long. It is white or yellowish white with a well developed prothoracic shield. Abdominal prologs are reduced and the crochets are short and stout, arranged in a single narrow ellipse. The larval period is 33-41 days. Before pupation, the exit hole is covered with thin webbing and a white silken cocoon is formed in which it pupates. The dark brown pupa is 12 mm long and the moth emerges in 6-10 days or in about a month depending on the climate. The moths are capable of emergence through 12 cm of water in the field. Female moth is bigger than male. The female has bright yellowish brown forewings with a clear single black spot and the anal end having tufts of yellowish hairs. The male is pale yellow and the spots on the forewings are not conspicuous. They are attracted to light, specially the females, which come from as far as five males. The female o male ratio is 2:1. Three to five generations or broods of the YSB occur in South India. October to December, characterized by cold weather, high humidity and low temperature are the crucial months.

Management strategies

- The removal and destruction of stubbles after the harvest decreases the carry over to next crop. Harvesting the plant close to the ground.
- Clipping off tip of seedling before transplanting reduced carryover of egg from nursery to main field. As the egg of stem borers are laid near the tip of leaf.
- Release T*richogramma japonicum, T. chilonis* an egg parasitoid@ 50,000 -1,00,000 adult/ ha.

- Avoid high dose fertilizer.
- Fields sowing more than 5 per cent dead hearts should be sprayed with 2.5 liters of Chlorpyriphos 20 EC in 250 liters of water/ha or 7.5 kg Phorate10G/ ha or Fipronil 3G @ 15 kg/ha or Cartap 4G @ 25 kg/ha.

2. Rice plant hopper (*Nilaparvata lugens* Stal.)

Host plant - Apart from rice, it infests *Cyperus rotundus* and *Panicum repens*.

Distribution - The insect is widely distributed in all rice growing countries and is Southern Japan to Oriental region.

Nature of damage – Both adults and nymphs have been noticed to suck cell sap from the leaves, turning them yellow. If the attack is during early stages of growth, entire plant may dry up. Under favourable conditions i.e. high humidity, optimum temperature, high nitrogen application, its population increases very rapidly. This is referred to as **"hopper burn"**, which gives brownish hue to the whole field. This pest also serves as vector of rice grassy stunt virus. Heavy population build-up of this pest, causes lodging of the crop and thus heavy yield loss.

Life cycle - Three development stages viz, egg – nymph and adult are found in the life cycle. The adult female inserts the whitish, transparent, slender, cylindrical and curved eggs inside the leaf sheath tissue in two rows each group having 9 – 33 eggs. Eggs are also laid on either side of the midrib of the leaf sheath or in the fleshy basal portion of the midrib. The eggs are closely packed and glued together at one end. Eggs measure about 1.0 mm long and 0.2 mm broad. A female lays as many as 212 eggs. Incubation period: 4-8 days. The nymph are small, creamy white with pale brown tinge (0.8 – 2.5 mm), at maturity turns yellowish to green. They pass through five instars and become adult in about 12-18 days (nymphal period). It is a brown planthopper measuring about 2.5 mm long in males and 3.3 mm in females. Brachypterous (short winged) female measures 3.4 mm long and macropterous (long winged) ones are 4.5 mm long. Prominent tibial spur is present on the hind leg. Adult longevity is 10 - 20 days. Adult Period Duration varies; June - October 18-24 days November - January 38 - 44 days February - April 18 - 35 days.

Management strategies

A. Preventive measures

- Use of resistant varieties Annanga, Aruna, Bharatidasan, Bhadra, Chandana, Chaitanya, Co42, Cotton Dora Sannalu (MTU 1010), Daya, Jyoti, Kanaka, Karthika, Krishnaveni, Manasarovar, Mekom, Nagarjuna, Neela Annanga, Pavizham, Pratibha, Rashmi, Remya, Sonasali, Vajram, and Vijetha, IR26, IR64, IR36, IR56 and IR72.

- Provide alleyways or pathways of 30 cm width after every 2-3 meters while transplanting.
- Remove excess seedlings and weeds from the field and bunds.
- Apply recommended dose of nitrogen based on soil fertility.
- Split application of fertilizers at the time of transplanting, tillering and panicle initiation stage (apply after weeding).
- Set up light traps at night to care should be taken not to place light traps near seed beds or fields.

B. Biological control

Predators Green mirid bug, *Cyrtorhinus lividipennis* Reuter (Hemiptera: Miridae) Carabid beetle *Ophionea nigrofasciata* (Schmidt-Goebel) (Coleoptera: Carabidae). Both the shiny black larvae and reddish-brown adults search the rice canopy for prey, consumes 3-5 hoppers per day. Larva of *Pseudogonatopus nudu* Lady beetle *Harmonia octamaculata* (F.) (Coleoptera: Coccinellidae). Larvae are black with dark yellow spots and have body horns dorsally and laterally. These are active during the day in the upper half of rice canopy. Larvae are more voracious than adults. They consume 5-10 prey a day. *Micraspis* sp. (Coleoptera: Coccinellidae) Adult is oval and brightly coloured shades of red. *Micraspis crocea* (Mulsant) (Coleoptera: Coccinellidae). Adults and larvae prey on small plant hoppers. Judicious and need based application of safer insecticides will conserve the natural enemies.

C. Control measures

Cultural practices

Alternate wetting and drying the field during peak infestation helps in managing the pest.

Draining out the standing water from the field 2-3 times also reduces the pest incidence.

Chemical control

Look for BPH at base of the plant 30 days after transplanting at weekly interval.

Spray at economic threshold level of 5-10 insects per hill, 100 ml of Imidacloprid 200SL or 1.4 liters of Monocrotophos 36SL in 250 liters of water per/ha or Thiamethoxom 25WG@ 0.2g/l or Phorate 10 G @ 12.5 kg/ha at nursery stage or 10 kg/ha in field crop or Carbofuran 3 G @ 33 kg/ha at nursery stage or 25 kg/ha in field crop.

Precautionary measures at the time of chemical spray

While spraying, nozzle should be directed at the basal portion of the plants. Repeat the application if hopper population persists beyond a week after application. At hopper burn stage, treat the affected spots along with their 3- 4 m periphery immediately as these spots harbour high population of the insect. Avoid use of synthetic pyrethroids (cypermethrin, deltamethrin) and quinalphos as they cause resurgence. Do not use same insecticide repeatedly when 2-3 applications are needed. Application with power sprayer is preferable. The quantity of insecticide required is 3 times more than the normal with knapsack sprayer. Ragged stunt virus disease: Symptoms of the disease are stunting, twisting and curling of the leaves especially flag leaf with ragged or serrated margins. Sometimes galls are present along with leaf. Grassy stunt virus disease: The symptoms include, stunting, profuse tillering, yellowish green leaves with numerous rusty spots which later form irregular blotches. There will not be any flowering and the plants present a bushy or grassy appearance.

3. Green leaf hopper (*Nephotettix nigropictus* stal. and *Nephotettix virescens* Distant)

Host plant - It is primarily a pest of rice in India but during off season it thrives on grasses

Distribution - The insect is widely distributed in all rice growing countries and is Southern Japan to Oriental region.

Nature of damage - Both nymphs and adults suck the sap from leaf sheaths and blades and cause browning of leaves or "hopper burn". In general its attack causes uniform yellowing of leaf from tip to the middle half of leaf. However, serious damage is inflicted when it transmits the virus diseases such as rice dwarf, rice yellow dwarf, rice transitory yellowing and rice Tungro. Plants infected with dwarf or stunt virus show symptoms of veins of leaves having a series of white dots on the midribs developing into parallel yellowish streaks. Leaves become chlorotic. The honey-dew excreted by the hoppers favour multiplication of sooty mould caused by *Capnodium* sp.

Life cycle - Three development stages viz, egg – nymph and adult are found in the life cycle. The adult female with its ovipositor makes a vertical incision in the leaf sheath and the yellowish eggs are laid in a row across the incision under the epidermis. The egg is about 1 mm long, 0.3 mm broad and white elongate or cigar shaped. A female lays upto 53 eggs. In a single row upto 24 eggs may be laid. Incubation period: 6-7 days. The nymph are transparent, white and shiny (0.9 -3.1 mm), at maturity turns yellowish to green. They pass through five instars and become adult in about 18 days (nymphal period).

Adult *N. nigropictus*: Male with two black spots extending to black distal portion of forewing, a black tinge along the anterior margin of pronotum and a sub-marginal black band on the crown of the head. In *N. virescens* has absence of the black sub-marginal band on the crown. In male the black spots do not extend upto the black distal portion of forewings. Further, the central tegminal marking usually does not touch the cleval suture. On the pronotum black tinge is absent. On an average the life cycle occupies 23 days. The female lives for about 55 days. Its peak period of occurrence in India is July to September and decreases markedly after a heavy rain.

Management strategies

As for Rice Plant Hopper (*Nilaparvata lugens* Stal.)

4. Gundy Bug (*Leptocorisa acuta* Fab.)

Host plant - Paddy Jowar, bajra, maize smaller millets and many grasses.

Distribution - The insect is widely distributed in all rice growing countries and all over of India.

Nature of damage - It is the most serious pest of the paddy crop. The nymph and adult suck juice from the developing grain in the milky stage, causing incompletely filled panicles or chaffy grains. Black or brown spots appear around the holes made by the bugs on which sooty mould may develop. Its attack can easily be recognized by unpleasant odour which comes from the field where it is present and due to which name 'gundhi' has been given. The insect is very common in paddy field from middle of August to middle November and causes 20 to 30% damage of crop.

Life cycle - Three development stages viz, egg – nymph and adult are found in the life cycle. The eggs are laid by the female bug in the month of July and August they are deposited in 2-3 stripes on the midrib of upper surfce of leaves. Each stripe contain 10- 20 eggs and bout 40 eggs are laid at the place. The eggs are blackish-brown small bead like and measuring 2 mm x 0.8 mm long, depending on the prevailing temperature. Incubation period is about 4-7 days. At the time hatching their eggs colour changes in to black. The nymph is slender greenish hatch out from eggs, start immediately sucking the plant juice. At the time of hatching it is about 18 mm long which grows up to 20 mm within 6 hours. There are five instars after which the nymph becomes adult with wings. The wings – pad may be seen in the 3rd nymphal instar. The nymphal period is about 14-20 days. The fully grown nymph is about 15 mm long. The adult bugs are slender, greenish yellow in colour. Head is small and provided with 4 segmented rostrum. The antennae are large than the body and red in colour. Meta thorax provided with glands which given unpleasant odour. The fore wings are hemeelytrate. The life cycle is completed in 21-31 days and several overlapping generations are found but on paddy 4-5 generations are completed. The bugs copulate after 12-14 days of emergence and female survived up to 55 days while male only 33 days.

Management strategies

- All weeds and grasses growing around the paddy field should be destroyed.
- The nymphs and adults are attracted in light traps.
- Resistance varieties should be preferred in infected areas such as Sona, Mugdha etc.

Natural enemies

Parasites: *Gryon* sp.

Predators: *Conocephalus longipennis*, Spiders.

Pathogen: *Beauveria bassiana*

Chemical control

- Dust 5 per cent Malathion or 5 per cent Carbaryl @ 25 kg/ha 15- 20 days after panicle emergence.
- Placing a rotten frog/ dry fish in the field during milky stage, the foul smell of these rotten material, attract bugs and keep busy on feeding carcass.

5. Rice leaf folder (*Cnaphalocrocis medinalis* Guen)

Host Plant - Paddy, Jowar, bajra, maize smaller millets and many grasses.

Distribution - The insect is widely distributed in all rice growing countries and all over of India.

Nature of damage - The larva fastens the edges of a tender leaf together and lives insides the rolled leaf by scraping the green tissues. Sometimes the tip of a leaf is drawn and fastened to the basal part of the leaf which gives it a rolled appearance. The infested leaves turn whitish and in cases of severe attack the whole field appears scorched and sickly. The infestation starts a month after transplantation and many continue up to the boot leaf stage of the crop. The growth of tillers gets affected considerably and results in a loss ranging from 5-60 per cent. It is a sporadic pest of importance.

Life cycle

Four development stages viz, egg, larva, pupa and adult are found in the life cycle. The female moth lays eggs flat oval yellowish singly or in pairs on the undersurface of tender leaves during evening hours or at night. Single female lays about 250-300 eggs in her life span. The hatching time varies from 4-7 days. The larva feeds inside the leaf fold and spins threads over and around itself as a protection. Later, it turns the edge of the leaf over in a fold and bend it down with silken threads, living safely within this fold. The full grown larva pale yellowish green and measures about 16-20 mm long. It moults four times and becomes full grown in 15-27 days. The pupation takes place either in rolled leaves or on the ground among fallen leaves. Before changing in to pupa, the larva shrinks and becomes punkish in colour. The pupa is reddish brown in colour and measures 12 mm. pupation period lasts in about 6-8 days. The adult moth is small, brownish orange in coloured with light brown wings having two distinct dark wavy lines on forewings and one line on hind wings, the outer margin characterized a dark brown to grey hand. Labial and maxillary palps fused together to form a snout. The pest is active from March to October-November. There are 5-6 generations in a year and total life cycle is complete 26-42 days depending upon environmental conditions.

Management strategies

- Remove grass weeds from bounds around paddy fields.
- Light trapping of adult helps to reduce pest population.
- Regular hand picking of rolled leaves minimize the attack of pest.
- Balance nitrogen should be given and if possible resistant varieties should be preferred.

Potential Natural Enemies

Parasites: *Trichogramma chilonis, Trichogramma japonicum, Temelucha* sp.

Predators: *Ophionea indica*, Spiders, *Andrallus spnidens*, *Nabis* sp.

- Release *Trichogramma japonicum* or *T. chilonis* @ 50,000 to 1,00,000 adult/ha.
- Spray insecticide at economic threshold level of 10 per cent damaged leaves 875 ml of Triazophos 40EC, 2-5 litres of chlorpyriphos 20EC 250 litres of water/ha.

6. Rice hispa (*Dicladispa armigera* Oliv.)

Host Plant - Paddy, sugarcane and grasses etc.

Distribution - The insect is widely distributed in all rice growing countries and In India, especially in Andhra Pradesh, West Bengal and Bihar.

Nature of damage -It often assumes serious proportions on young paddy crop. The damage varies from 25-65 per cent. The beetle itself feeds on the green matter of the tender leaves producing the characteristic narrow white lines on them. The grubs mine into the leaf tissue and eat up the leaf contents; the presence of these grubs in the leaf tissue is indicated by the peculiar blister spots towards the tip of the leaves. In cases of severe attack the leaves turn brown and wither away presenting a sickly with appearance.

Life cycle -Four developmental stages viz. egg, larva (grub), pupa and adult are found in its life cycle. The female lays its eggs inside the tender leaves, generally towards the leaf tip. Eggs are minute slits yellowish in colour. A female bettle on an average lays 55 eggs and the incubation period is 4-5 days. The small grubs soon after hatching are quite active and begin to feed on the leaf tissue inside the leaf mine. They are yellowish in colour with become flattened shape. There are four larval instars and total period 7-12 days. At the end of the last instar when the larva is full grown, it pupates in the mine itself. The pupa is radish-brown in colour and measuring 5 x 4.5 mm long. The pupation period varies from 3-5 days. The adult beetle is steel blue to black in colour and measures about 4.5 – 5 mm long. Its bear a short single spine on elytera and paired four pronged and single spines on thorax. The adult beetle soon after emergence begins to feed and starts eggs laying after a week. The life cycle is complete in 14-22 days. There may be about six generations in one year.

Management strategies

- Clipping of leaf tip before transplantation.
- When nursery bed is flooded, the beetles float that can be collected at a corner of nursery and destroyed.

Potential Natural Enemies

Parasites: *Bracon hispa.*

Predators: Spiders.

Management strategies

- Application of Phorate 10 G in nursery minimises infestation.
- Spray at economic threshold level (one adult or 1-2 damaged leaves per hill) with or Chlorpyriphos 20EC in 250 liters of water/ha.

7. Rice Grasshoppers [(i) *Hieroglyphus banian* Fab. (ii)*H. nigrorepletus* Bolivier]

Host Plant - Paddy, Jowar, bajra, maize, sugarcane, smaller millets and many grasses.

Distribution - This is a sporadic pest of paddy distributed all over the world. In India it is found almost area and around the year.

Nature of damage - They are polyphagous pest, nymphs and adult generally feed on grasses over the bunds of rice fields, before attacking the crop. The leaves are completely eaten, leaving the midrib and stalk. In the earhead stage they, as adults, attack the ears or nibble at the tender florets or gnaw into the base of the stalk, leading to the formation of 'white ears'.

Life cycle: Three development stages viz, egg – nymph and adult are found in the life cycle. Eggs are laid in soil at a depth of about 5 cm during October-November and the eggs are covered by a gelatinous substance. Generally, breeding starts with the onset of good monsoon showers. A female may lay about 100-150 eggs in her whole life span. The eggs are yellow in colour and resemble to rice grain. It is about 78 mm long and of 1 mm girth. They hatch 12-14 days in summer and 4 weeks during winter and eggs lie dormant for a period of 8- 9 months. At this stage a suitable amount of soil moisture is most essential for the proper development of eggs. If this prerequisite is provided by timely rainfall the eggs hatch into a vermiform larvae which wriggle their way out of the soil sand, after casting of a kind of membranous convening in which they are enveloped, they enter the hopper stage of their life. Some after emergence from the soil and casting off the membranous covering, the young ones now called nymphs begin to feed and grow. It is light yellow in colour measuring about 6 mm long and devoid of wings. They start feeding on grasses and can be seen jumping from one place to another. This nymphal period lasts upto the end of August or early September when the insects attain the adult stage. This period is 2.5-3 months and its has found 6-7 instar. After second moult the colour changes to light green and wing pads are visible. The adult hoppers are yellowish-green in colour. *H. nigrorpletus* is larger than *H. banian*. The male and female of the former are 37 mm and 75 mm long while of the later 25 mm and 37 mm long, respectively. In case of *H. banian* there are only 2-3 black marking but in *H. nigrorepletus* there is a reticulum of black lines and possesses U shaped black spot on the lateral sides of pronotum. After becoming adults the insects take 7-20 days to attain sexual maturity and after the first mating they require a period of about seven days to lay the first batch of eggs. There is only one generation in a year. The life of adult lasts about two months.

Management strategies

- The eggs can be destroyed by digging them.
- The majority of eggs are laid in the soil which is generally not cultivated areas or bunds and mounds within the cultivated portion. Due to this peculiarity, it becomes necessary to reduce the uncultivated area as much as possible. Reclamation of cultivable wasteland should reduce the intensity of the problem posed by the grasshopper. Particular attention should be paid to level down all the stray mounds in the fields.
- It has shown beyond doubt that mere mechanical disturbance of the soil wherein the grasshopper eggs have been laid, immensely reduces the chances of successful emergence of grasshopper nymphs. Hence, it is advisable that before the monsoon begins, the bunds are scrapped to a depth of about 15 cm and reformed.
- After the onset of the monsoon a period of about one week has to lapse before the hopper nymphs emerge out of the soil. This period should be utilized for treating the egg-infested areas with persistent insecticides which would kill the nymphs as soon as they emerge. Dust 5 per cent Malathion or 5 per cent Carbaryl @ 25 kg/ha.
- Spraying the crop with 0.1% neem seed kernel powder found very promising and it protects the crop for 3 weeks.

8. Paddy Root Weevil (*Echinocnemus oryzae* Marshall)

Host Plant - Paddy and some grasses etc.

Distribution - The insect was first reported in 1924 from Gunter in Andhra Pradesh. It occurs in Tamil Nadu, Kerala and Punjab etc.

Nature of damage - The larvae (grubs) are aquatic and feed on the rootlets of paddy, grasses and *Fimbristylis tenera*. The attacked plants become stunted and do not put forth tillers. It causes a loss of 5-10 per cent in rice. The crop transplanted in July is more heavily attacked than the one transplanted in August.

Life cycle - Four developmental stages viz. egg, larva (grub), pupa and adult are found in its life cycle. The total life cycle from egg to adult takes 42 to 54 days.

Management: Apply 20 kg of Carbofuran 3G/ha in standing water.

9. Rice Caseworm (*Nymphula depunctalis* Guenee)

Host Plant - Paddy and some grasses like *Eragrostis* sp., *Panicum* sp. etc.

Distribution - The insect is distributed in India, Pakistan, Sri Lanka, Australia, and Philippines. In India occurs in mostly all rice grower state.

Nature of damage - Early stages of the crop are damaged by the pest the larva cuts the leaf blades into short lengths and constructs a tubular case inside which it remains and feeds on the foliage scraping the green matter in patches leaving characteristics white marks. The larva mostly feeds on the undersurface of leaves leaving the upper epidermis alone insect which appears white. Rice hills infested severely become stunted and often completely get killed. The damage is severe in the early stage of transplanted crop.

Life cycle - Four developmental stages viz. egg, larva, pupa and adult are found in its life cycle. The female lays the eggs singly on the leaves of rice and grasses. The eggs are flat oval and white in colour which turns to brown at the time of hatching. Single female lays about 50-60 eggs in her life span. The hatching time varies from 2-6 days. The young larva feeds on inside a tubular case made up of cut leaf bits and carries it as it moves. With each moult, as it grows, the leaf-case is also changed. The full grown larva light green in colour with light brownish-orange head and measure about 15 mm long. The larva is semi-aquatic in habit. It has filamentous gills on the sides of its body and the necessary oxygen is obtained from the water in the leaf-case. This water in the leaf-case is often replenished with fresh water by the larva coming to the surface of water in the field. Total larval period is 14-20 days. The pupation takes place inside the leaf case which remains attached to the base of the tillers. The pupa is reddish brown in colour and measuring 12 mm long. Pupal period lasts in about 4-7 days and are capable of emerging through water. The adult moth is whitish in colour with faint yellow wings, the wings with many fine dark spots and one dark lines, forming an irregular pattern. It measures about 12 mm long with wing expanse of 25 mm. The total life-cycle occupies about 19-37 days.

Management strategies

- Dislodging the leaf-cases from the plants by passing a rope and draining the water.
- Drain-of the water from the field to kill the floating larvae.
- The addition of a small quantity of kerosene to from a thin film on water kills the larvae
- The insect can be controlled by spraying the crop with 1.0 litre of in 250 litres of water/ha.

Potential Natural Enemies

Parasites: *Dacnusta* sp.

Predators: Spiders

10. Rice Gall Midge (*Orseolia oryzae* Wood-Mason)

Host Plant - Paddy and some grasses like *Cynodon dactylon*, *Eleusine indica Panicum* sp. etc.

Distribution - The insect is distributed in India, Pakistan, Sri Lanka, Burma, China and Sudan. In India occurs in mostly all rice grower state.

Nature of damage - The symptom of attack is the formation of the leaves feed on the growing points producing characteristic gall in the plant transforming regular tillers into tubular galls which dry off without bearing panicles. The leaf sheath changes into a hollow, dirty white or pale green, pink or purple, long cylindrical tubular gall at its tip. The manifestation of injury is in the form of "**silver shoot**" or "onion shoot". The pest starts infesting the plants in seed bed and can continue to do so until booting stage. Heavy infestation may result in loss up to 50 per cent.

Life cycle - There are four developmental stages viz. egg, larva (maggot), pupa and adult are found in its life cycle. The female lays 100 to 300 eggs singly or in groups of 2-6 on the hairs of ligules of the rice leaf or on the leaf sheath just below or above the position of ligules. Occasionally they may be laid on standing water. The egg is reddish, elongate, tubular and 0.55 mm long with rounded ends. The incubation period is from 3 – 4 days. The first instar larva hatches out from the egg on the leaf moves down in about 6 – 12 hours to the shoot apex without boring into plant tissue. It feeds at the base of the apical meristem. Due to larval feeding the apical meristem gets suppressed and induces formation of radial ridges from the innermost leaf primordium, just above the level of the posterior end of the larva, which is followed by the elongation of the leaf sheath. If more than one larva reaches the shoot apex only one will survive to maturity and if the first or second instar larva dies during the course of its development, the apical meristem gets reactivated and resumes development into a normal shoot. This phenomenon does not occur if the larva dies in its later stages of development. It appears that the development of the radial ridges can be attributed to diversion of nutrients from the apical meristem to the larva. The subsequent gall elongation is possibly due to substances produced by the first instar larva and prepupa. Normal development of the larva proceeds, if the first instar larva infests only the growing shoot apex and development is inhibited if it attacks an inactive auxiliary shoot apex. In cases of severe outbreak, through almost all shoot apices are infested, due to larval dormancy there is a staggered development of gall and adult emergence. The pest carry over to the next season is accomplished by the larvae remaining dormant in active auxiliary shoot apices of host plants. The full grown larva is 3 mm long and its pale red in colour. Total larval period is about 20 - 25 days. Pupation takes place at the base of gall. The pupa with the help of several rows of backwardly pointed sub equal abdominal spines wriggles up the gall of the tip and its makes a hole with spines at the anterior end and projects half way out. Then the adult emerges, usually during the night. Total period is about 20-23 days. The adult fly is yellowish-brown in colour and mosquito like structure. It measures about 3-3.5 mm long. They are active at night. The total life-cycle occupies about 42-46 days.

Management strategies

Potential Natural Enemies

Parasites: Larval parasite *Platygaster oryzae, Neanastatus oryzae.*

Predators: *Ophionea indica, Cyrtorhinus lividipennis*, Spiders.

Resistant variety : BPT 16

- Application of three to four round Fipronil 3G @ 15 kg/ha or Cartap 4G @ 25 kg/ha during planting to tillering stages of the crop.
- At panicle initiation to booting or flowering and after, spray Chlorpyriphos @ 0.5kg/ha.

11. Armyworm (*Mythimna separata* Walker)

Host Plant - Paddy, sugarcane, sorghum, maize, wheat, oat, pea, *Panicum* sp, grasses etc.

Distribution - The insect is distributed in India, Pakistan, Sri Lanka, Burma, Thailand and Philippines. In India occurs in mostly all over state.

Nature of damage – The larvae moves in bands and destroys all the crops that come in their way, especially cereal crops as mentioned above.

Life cycle: Four developmental stages viz. egg, larva, pupa and adult are found in its life cycle. The female lays the eggs shining white spherical with fine reticulations. They are laid in rows or in masses of 20 – 76 between overlapping leaf sheaths of sugarcane or on rolled leaves or under leaf base of rice. The hatching time varies from 4-6 days. The full grown larva is dirty pale brown or dark with a median dark brown line and two dark brown and one white lateral strips. The head is grayish brown. Total larval period is 22-24 days. The pupation takes place in the soil in an earthen cell or sometimes inside leaf sheaths on plants. The pupa is redish brown in colour and measuring 12 mm long. Pupal period lasts in about 7-9 days and are capable of emerging through water. The adult moth is pale brown with dark specks in colour with hindwings pale with brown tings. It measures about 25 mm long with wing expanse of 40-50 mm. Fore wings are dirty white in colour having dirty brown spots while hind wings are white. The terminal end of abdomen bears a tuft of hairs. The total life-cycle occupies about 35-40 days.

Management strategies

- Moths may be destroyed after attracting them in light-traps.
- Hand-picking of larvae may also reduce the population of caterpillars.
- Flooding of field may kill the caterpillars.

12. Climbing cutworm [*Agrotis ipsilon* (Hufn)]

Host Plant - Maize, sorghum, wheat, oat, pea, potato, tobacco, bhindi etc.

Distribution - The insect is distributed in European countries, Pakistan, America, Africa, Thailand and Philippines. In India occurs in mostly all over state particularly UP, Bihar Madhya Pradesh, Rajasthan etc.

Nature of damage - In India, winter is generally lean season for insect-pest activity, but the cutworms belong to that small group of insects, the destructives activity of which is more marked in the rabi-crops. They are known as cutworms because they cut and fell down to the ground either the whole plants like maize their twigs. The damage caused by the larvae which hide during the day and come out in the evening to do the damage. First of all they feed on the epidermis of the fallen leaves or the green leaves touching the ground. Latter the caterpillars cut the leaf or shoot or plants just above the ground level buried in the soil. The attack of cutworm in maize is confined from November to February and causes a very serious loss.

Life cycle - Four developmental stages viz. egg, larva, pupa and adult are found in its life cycle. The female moths of the cutworms are strong fliers. They come to the plains in winter from hills and return in summer to hills again. After mating, the females lay eggs mostly on the underside of the leaves or on the soil near the root base. The eggs are laid singly or in clusters of up to 30 eggs. A female generally lay 300-350 eggs but can lay thousands of eggs. The eggs are round yellowish in colour and measuring 1 mm long. The hatching period of eggs varies from 2-6 days but during severe cold 12-15 days. The full grown larvae feed on their egg-shell as their first meal and fall to the ground at the slightest disturbance. They are light bluish green in colour and come out at night. The full grown larva is about 40-45 mm in length, plump smooth and flattened in look and dull grey brown in colour. The larva becomes full grown in 21-36 days. The pupation takes place in the soil. The full grown larva enters the soil much deeper than during its usual activity constructs the earthen chamber, the sides of which are made smooth and hard by an exudation from the larval mouth. The pupae are reddish brown in colour and about 20 mm in length and 7 mm in breadth. The pupal period is about 10-30 days. The adult moth is brown in colour measuring 25 mm in length and 7 mm in breadth with a wing expanse of 40-50 mm. the male has bi-pectinate antennae while female has filiform. Fore wings are dirty white in color having dirty brown spot while hind wings are white. The terminal end of abdomen bear tuft of hairs. This is active from October to March in plain and after which most of them migrate to hills. The life cycle is completed in 35-84 days and generally two generations are found in plains and in hills four generations occur during summer.

Management strategies

- Same as that for army worms.

Natural enemies

- Larval parasite of the pest: *Braconid* wasp, *Apanteles* sp. and *Microbracon* sp.

13. Termites (*Odontotermes obesus* Ramb.)

Host Plant - This is polyphagous pest and feeds on every variety of agricultural crops. Among the crops grown wheat, maize, sorghum, wheat, oat, pea, potato etc.

Distribution - The insect is prevalent in all the warmer parts of the world but commonly found in the tropical regions.

Nature of damage - This is a polyphagous pest but mainly feeds on wheat and maize. Termites attack both maize stalks and roots. As a result of their attack, the young maize even the shoot of older maize dry up and finally the growth is adversely affected. The maize roots are eaten up from the cut ends and are rendered hollow. In unirrigated areas the young wheat crop is damaged badly, as they destroy the roots of seedlings ad finally the plants are dried off completely. They destroy the crop at night and remain hidden in their nest during day time. The attacked plants can easily be distinguished by irregular cuttings and adhering of soil particles. The fruits trees are also attacked and on branches and stem the soil channels are formed under which they live.

Life history - Three developmental stages viz., egg, nymph and adult are found. They are the social insect and live in colony. There is polymorphism in adult stage and king, queen soldiers, workers are found. The members of the colony may be divided into (i) reproductive forms (queen, king, complimentary form and colonizing form). (ii) Sterile form (workers and soldiers).

The winged reproductive's comes out in swarms from the nest generally in the beginning of monsoon. Swarming usually takes place in the day time and most of the individual of the swarms are destroyed by birds etc. the survivals mate, shed their wings and burrow in the ground to form a new colony of which they become the king and queen. Their flight is known as nuptial flight. The queen lays first batch of 10-130 eggs about a week of swarming, but later on in large numbers about 30,000-40,000 eggs per day throughout her life time, which may be five years. Eggs are small kidney-shaped and yellowish in colour. Egg is 0.5 mm long and incubation period about in a week. The newly hatched nymphs are yellowish white in colour and about 1 mm long. They eat for some time the excreta of the king and queen and latter search the food. Nymphs develop into different castes in 6-13 months after 4-10 moulting.

1. Reproductive form

(a) **Queen:** The queen is the largest individual ranging 6-8 cm in length and 1 cm in thickness. She is mother of the colony and lives in a specially prepared royal chamber which is situated in the centre of the nest at a depth of 1-6 feet below the ground surface. The queen is wingless, creamy white in colour and her abdomen is marked with transverse dark brown bands.

(b) **King:** The winged male which remains with queen after nuptial flight is known as king. There is generally a single king in each colony which is smaller than queen and remains with her in royal chamber.

(c) **Complimentary form:** There are apterous or brachypetrous forms of both the sexes which maintain the numerical strength of the colony in absence of macropterous forms. In a single colony their number may be in hundred.

(d) **Colonizing form:** They are generally produced in rainy season and attracted by light. They are brownish in colour with two pairs of slender, dark brown, long narrowed wings which are used for nuptial flight and after which they are she. These form emerge in millions from the parent colony during monsoon and after finding their mate they craw into the soil and start the new colony. These are thus the kings and the queens of future colonies.

2. Sterile form

(a) **Worker:** The worker constitutes the main labour force of the colony and he is about 80-90% of the whole colony. It is about 6-8 mm long, dirty white in colour with brown head and eyes is small or absent. Mouth part is fairly strong and used as the working tools for all types of odd job.

(b) **Soldier:** The soldier is slightly bigger than worker. Mouth part is well developed and used for very active defense and offence. The main function is to protect the colony but also help the worker is keeping the nest neat and clean by removing the dead and sickly members of the colony. Worker and soldier contain sterile individual of both sexes.

Management strategies

- The removal of dead or decaying matter or dry stubbles from the field useful because they attract the termite.
- The use of partially decomposed manure should be avoided.
- Irrigation water with crude oil emulsion may be used to avoid the termite attack.
- Mixing of any one of the following insecticide before sowing in soil has been found very effective.

14. (Plant lice (Aphid) *Rhopalosiphum maidis* Fitch)

Host plant - Wheat, maize, sorghum, bajra etc.

Distribution - In India it is found in all the states specially in Punjab, U. P., M. P., Maharashtra, Bihar, Gujarat, Karnataka and Rajasthan.

Nature of damage - Winged adults appear in the first week of January and by mid January. Both nymphs and adults suck the sap from the tender portion of the plants. They are confined to the unfurled leaves of the central whorl of wheat plants. Generally they are less abundant on wheat. Heavy infestation causes yellowish patches on leaves and ultimate drying of the plants. It transmits *chirke* virus of certain wheat varieties. It also vector of the grassy-shoot disease and mosaic in sugarcane.

Life cycle - Different type of life-cycle has been found in plain and hill regions. In plains, the insect does not lay eggs hence only nymph and adult are found. In plains, the female is viviparous it do not lay eggs but produce many nymphs. The nymphs are produced by following ways. (i) After mating the males (ii) Without mating the males, parthenogentically. The nymphs become adult within 9-10 days and start producing apterous offspring. In this way it increase great in number and spread over a big area in a very short time. There are several generation during the cold season. In hills aphid is remarkable on account of their peculiar mode of development and the polymorphism exhibited in different generations of the same species. The following types of individuals are present in the life cycle of migratory aphid.

(i) **Fundatrices:** It is apterous, viviparous parthenogenetic female which emerge in spring from the overwintered egg. Only one generation is found of this group.

(ii) **Fundatrigeniae:** There is also apterous, viviparous parthenogenetic female and are the progeny of fundatrices. It lives on the primary host and complete 3 generations.

(iii) **Migratory:** These types usually develop in the second third or later generations of fundatrigeniae and consist of winged parthenogenetic, viviparous females. They develop on the primary host and subsequently fly to the secondary host.

(iv) **Alienicolae:** Parthenogenetic, viviparous females developing for the most part on the secondary host. They often differ markedly from the fundatrices and migrants. They are similar to fundatrigeniae but differ them by comprising both apterous and winged forms.

(v) **Sexuparae:** Parthenogenetic, viviparous females which develop on the secondary host and later migrate to the primary host at the end of summer.

(vi) **Sexuales:** This is the progeny of sexuparae produced in the primary host. They consist of sexually reproducing male and female. The female laid on the primary host after copulating the male which passes winter as such and hatch out in following spring and fundatrices female is produced.

Adult: The aphid is small generally 2 mm in size, green sucking insect. He is provided with a pair of small tubular structure projecting out from dorsal surface of the posterior region of the body known as cornicles or siphon or honey tube. Two pairs of transparent wings are found in which costa and sub costa veins are absent. Winged adults are comparatively few in number in a population.

Management strategies

Spray of 0.025% parathion or 0.05% carbophenthion or 0.025% methyl demeton will control the pest.

15. Pyrilla (*Pyrilla perpusilla* Wlk.)

Host Plant - Wheat, jowar (sorghum), bajra, barley, maize and other plant s of poaceae family.

Distribution - Generally it is found all over India. But it is more prevalent in U.P, Bihar, Punjab and Maharashtra.

Nature of damage - Both the nymphs and adults suck the cell sp from under surface of the leaves and attacked leaves become yellowing and withering of leaves. The honey dew excreted by the pest attracts sooty mould which interferes the proper functioning of leaves and thus renders them unfit even for the cattle feed. Growth of the plant is suppressed and when terminal buds are affected, lateral sprouting may take place.

Life cycle - There are three stages in its life cycle viz., egg, nymph and adult. The female lays in cluster of 2-60 generally on the underside of the leaves. The eggs are covered with with cottony filaments from the anal tuft of the female and are bright green yellow or dark brown in colour, oval in shape and 1 mm in diameter. The eggs hatch in 7-12 days during summer and 3-4 weeks during October –November. A female lays about 750 eggs. The nymph is pale brown in colour and 1.2 mm long with two long filaments in the anal region and is covered with white wax. They start sucking the sap of the leaves, grow in size and change into adults after 5-6 instars usually, there are 5 nymphal instars. The total duration of nymphal stage may vary from 4-6 weeks in summer and 15-22 weeks during winter months. The adult hopper is straw coloured measuring 9 mm long with wing expansion of 23 mm. There are two pairs wings folded roof shaped on the head extended anterior from a beak like structure snout. The head is provided white back large eyes and 3 segmented antennae the females can be distinguished from the males by the presence of pir of pad on the posterior end of the body. The females survived 6-8 weeks, white male only 4-6 weeks. The life cycle from egg laying to adult stage is generally completed in 42-63 days and 4 generation in a year. The first 2 broods are clear cut but the third and fourth overlap each other. The first generation are eggs laid in first week of April, 2nd generation from last June to middle of July, 3rd generation in the middle of August and fourth generation are laid during November and December after which adults usually die due to cold.

Management

- Burn all trash after harvesting the crops.
- Remove all the leaves bearing egg-clusters during March –April.
- Resistant varieties are growing.

  Natural enemies: Egg parasite – *Cheiloneurus pyrillae, Tetrasticus pyrillae*. Predator – (*Brumus saturalis*).
- Dustings have been found very effective in controlling the pest such as Carbaryl 10% @ 30 kg/ha.

16. Rice skipper (*Pelopidas mathias* Fb.)

Host Plant - Rice, maize, sorghum, bajra, sugarcane etc.

Distribution - This is found in all rice grown area in the world. In i.e. Tamil Nadu, Andhra Pradesh, Karnataka, Bihar, U.P., Jharkhand and Orissa.

Nature of damage - The larva has longitudinally folds the leaves making the edges stick together by means of silken threads and feeds on the green matter from inside. Sometimes a few leaves of tillers are closely webbed together and the larva feeds on the leaves resulting in stunting of hills in rice crop.

Life cycle - The life cycle of this pest passes through the four stages viz. egg, larva, pupa and adult. The adult female lays eggs singly on leaves in 2 or 4 days. Eggs are spherical flat based shape and creamy white in colour. One female lays about 30-95 eggs in its life cycle. Incubation period is about 3-6 days. Newly hatched larva feed on leaf sheath and full grown larva folds the leaves. Full grown larva measures about 35 mm long, pale green with yellowish-white lines across the back and laterally a whitish line on either side. The pale green head has two vertical red streaks. The larval period is about 13-39 days depending on climatic conditions. The pupa has pale green in colour and white longitudinal lines on it and is attached to the leaf blade by its caudal extremity supported by a silk girdle. Pupal period is about 7-30 days. The skipper is brown with whitish specks on each of the forewings. The insect pass through 4 overlapping generation during August-November and thereafter the larvae hibernate and emerge as adults in sprig. The total life cycle may be about 6 weeks.

Management strategies

- Drain the water.
- Spray chlorpyriphos 20 EC 1.25 l/ha

17. Sorghum stem borer *(Chilo partellus* Sinhoe*)*

Host - Sorghum, rice, maize, Johnson grass, Sudan grass etc.

Distribution - The insect has a wide distribution in India, Pakistan, Sri Lanka etc. In India Assam, West Bengal, Bihar, Odisha, U.P., etc.

Nature of damage - This is primarily pest of sorghum. The damage to young plants is far more serious it causes 'dead hearts'. Plants are not killed but suffer in vigour and develop weak heads. It causes damage up to 80 per cent.

Life cycle - The life cycle of this pest passes through the four stages viz. egg, larva, pupa and adult. The eggs are scale-like, flattish, oval, overlapping and laid in batches. A female lays 300 eggs usually on the underside of leaves near midrib and less frequently on stalk. Incubation period is about 7 days. Newly hatch larvae bite their way into the stem feeding on the internal tissue and killing the central shoot in young plants. The midribs of sorghum plants are often mined by the newly hatched larvae. The larva is cylindrical, yellowish brown with a brown head and prothoracic shield and measures about 25 mm long. The larva cuts a hole near one end of its burrow and plugs it with silky material. Larval period about 28-35 days but it extends to 193 days during winter as the larva remains in hibernation. Pupation takes place in specially constructed chamber in the stem. Pupa is creamy yellow and turns reddish brown in colour after about a day. The pupa is about 10-13 mm in length. The male pupa is narrow and smaller than female pupa. The pupal period about 7 days and moth emerges from the stem through a hole made by the larva at the time of pupation. The moth is medium sized and straw in colour. The fore wings are pale straw coloured while the hind wings possess a double row of black spots along their outer margins. The tip of the abdomen in female is dilated and tufted with hairs. While in male is pointed and devoid of tuft. The male died after copulation while female after 2-3 days of egg laying. They are nocturnal in habit and during the day remain hidden under dried leaves, clods of earth etc. Total life cycle is completed in 7 weeks and in winter about 83-210 days. There are 4-5 generation in a year.

Management strategies

Natural enemies: Egg parasite *Trichogramma minutum*

Larval parasite : *Apanteles colemani, A. flavipes, Bracon chinensis* etc.

- A higher seed rate is adopted and in the early stages affected plants are pulled out and destroyed.
- After harvest the stubbles should be removed and destroyed.
- Collection destruction of all the dead hearts and attracting the moth in light-trap proved effective control.
- Placing of Carbofuran 3G granules in the central whole of the plant on 25th and 35th day of sowing at 8 and 10 kg/ha.
- Application of 2-3 rounds of sprays of carbaryl 0.1% at 15 day intervals from a month after sowing minimizes the incidence of pest.

18. Cob caterpillar *(Cryptoblabes angustipennella* Hmpsn. *Cryptoblabes gnidiella)*

Host - Maize, sorghum, bajra etc.

Distribution - Sri Lanka and India. In India mostly maize growing area such as Uttar Pradesh, Bihar, Haryana and Rajasthan.

Nature of damage - The larvae infest the ears of sorghum and bajara, and cob of maize. The caterpillar remains near the rachis. Inside a thin web and webs together adjacent grain and feed on them.

Life cycle - Four development stages viz. egg, larva, pupa and adult are found in its life cycle. The female moth lays eggs flattened, somewhat triangular oval and creamy white on the lemma of newly opened flowers and rarely on the glumes and rachis. A female lays about 14 eggs in three days. Incubation period is 3 days. The larva is narrow, slender and grayish in colour with a dark brown mid-dorsal line and lateral lines. It measures about 12 mm long. The larva becomes full grown larvae in 19-22 days. The pupa is a thin silken cocoon and the adult emerges in 7-12 days. Adult is 7 mm across wing with forewings dark grey in colour. Total life cycle is 31-43 days. It occurs during summer and winter season.

Management strategies

- Placing Carbofuran 3G granules in the central whole of the plant on 25th and 35th day of sowing at 8 and 10 kg/ha.
- Application of 2-3 rounds of sprays of Carbaryl 0.1% at 15 day intervals from a month after sowing minimizes the incidence of pest.

19. White grub *(Holotrichia consanguinea* Blanch*)*

Host - This pest is polyphagous pest and feed on almost all the *kharif* crops like maize, jowar, shorghum etc.

Distribution - This is worldwide importance as noxious crop pest and cosmopolitan in occurrence. White grub has attained a status of major pest of a variety of rainy season crops in India. i.e. Rajasthan, Gujarat, Maharashtra, Karnataka, T.N., Bihar U.P. Jharkhand and Odisha.

Nature of damage - The damage caused be grubs and adult beetle former feeds on fine root lets, nodules and main root, whereas the adults feed on shrubs and trees growing nearby the cultivated fields. The plants damaged by the grub gives a wilted appearance and finally dries up, while in case of beetles the attacked plants get defoliated. The damage caused by this pest estimated to the tune of 40-80%.

Life cycle: The life cycle of this pest passes through the four stages viz. egg, grub, pupa and adult. The beetle emerge from the soil after dusk (7.30-8.00 pm) followed by good pre-monsoon rains, which may occur from April to June. After feeding and mating the beetle get back to the soil in early morning hours and lay eggs. This is generally laid singly in loose soil or in an earthen cell inside the soil upto the depth of 10cm. The eggs are oval, creamy white when fresh and later turn to brown in colour. About 30-120 eggs are laid by single female and hatching period about 7-14 days. Newly hatch grub in creamy white, measure 10-12 mm in length and 2-2 mm width, and feed on organic matter. There are three larval instars and full grown grub is curved 'C' shaped and dirty white in colour. Head is dark brown with strong mandibles and prominent thoracic legs. The larval period for the first and second instar 9-12 days and 3-4weeks, respectively, while n 3rd instar about 6 weeks. Thus total larval period varies from 72-82 days. The full grown grubs move down deeper in the soil in search of moisture and for pupation. Its constructs an earthen cell in which t passes a quiescent or prepupal stage which lasts for 1-6 weeks. Pupa is light yellow and extremely tender, but as it grows older it turn brown n colour. This period varies from 10-27 days depending upon the type of soil. Newly formed beetle is cream in colour with soft white elytra with the lapse of time the colour changes to brown and elytra gets hardened. Generally the adults are lamellate and males being smaller than females. Beetles remain within the earthen cell and do not come out of soil until the onset of rains. Total life cycle is completed from 90-108 days and all the known species in India have one generation in a year.

Management strategies

- The use of light traps is the successful method for the collection of beetles during emergence in the night.
- The repeated ploughings, preferably soon after the summer rains, help in exposing the various stages of white grubs to their natural enemies.
- Biological control or natural control such as *Bacillus thuringiensiis, B. popillae* (bacteria) and *Beuveria bassiana, Aspergillus parasiticus* (fungus).
- Adult beetle controlled by the solid application i.e. Phorate 10G @ 25 kg/ha or Carbofuran 3G @ 35 kg/ha.
- Spray of Chlorpyriphos 20EC @ 5 liters per ha offers the economic and effective control of grubs.

20. Rodents in cereal crops

Rodents are one of the most important non - insects pests of agricultural crops, particularly *kharif* and *rabi* cereal.

Types of rodents

- Lesser bandicoot rat: *Bandicota bengalensis*
- Field mouse: *Mus booduga*
- Indian gerbil: *Tatera Indica*
- Soft furred field rat : *Rattus meltada*

Damage at different stages - In India, rodents have been estimated to cause 5 to 10% losses in cereal crops. Among the field crops, rice and wheat are the most vulnerable crop to rodents. In addition to tiller cutting, they also hoard ripened panicles inside their burrows.

Nursery in rice - The nurseries are drained out and the rodents run freely inside the bed spoiling all germinated seed. Later, they also cut the seedlings 1-2 inches above the water level.

Main field - Sometimes the rodents pull out the transplanted seedlings in paddy and create gaps in the main field. Generally, their activity is confined to inside field leaving 2-4 meters on all sides of the field. In the initial stage, damage appears in patches and after some time, all these small patches become into one big patch. Damage increases with the onset of panicle initiation and continues up to panicle emergence.

Management strategies

- In Local traps called ' butta' are extensively used for the control of rodents in cereal. These traps provide fairly good results when applied after chemical control operation.
- However when directly used, trapping will be costly affair and one cannot manage entire population over large areas.
- Moreover, at certain crop stages, like primordial formation, rodents are not attracted towards traps.

Natural Smoke

- The main principle involved in this operation is simply filling the burrows with smoke, which causes suffocation to rodents ultimately leading to their death.
- The smoke liberated by burning rice straw mainly contains carbondioxide.

Chemical Control Fumigation

- The fumigants like Aluminum phosphide is effective and widely used for the control of field rodents living in burrows.
- The control of rodents using rodenticides is the more common way
  - Acute rodenticides (Single dose and quick acting), Eg : Zinc phosphide.
  - Chronic rodenticides (Multi dose and slow acting), Eg : Warfarin, Bromodiolone.

1) Acute Rodenticides: Among the acute rodenticides, Zinc phosphide and Barium carbonate are registered for use.

Action plan for Rodent control

Day-1	Identify live burrows and place 20 g of pre-bait Material inside the burrow.
Day-3	Place 10 g Zinc phosphide poison bait inside the Burrow.
Day-4	Collect dead rats and bury them. Close all the Burrows.
Day-5	Eliminate the residual population through trapping or Burrow fumigation with burrow fumigator. Treat the opened burrows with aluminum phosphide 2 pellets per burrow.
Day-13	In dry black soils fumigation will not give results. Hence apply Bromodiolone 1 cake per burrow. Repeat Bromodiolone baiting.

Advantages of Zinc Phosphide

1. Quick killing
2. Small quantity of chemical is required
3. Single feeding
4. Population can be brought down immediately.

Disadvantages of Zinc Phosphide

1. Necessity of prebaiting,
2. Low killing around 40-50 %,
3. Induce bait shyness.
4. Toxic to non target species,
5. Chances of secondary poisoning are more.

2) Chronic Rodenticides

- In order to overcome limitations and hazardous nature of acute rodenticides, lengthy baiting programme and possibility of resistance, new series of rodenticides have been developed and known as single dose anti-coagulants or second generation anti-coagulants. These rodenticides (Bromodiolone) combine better qualities of acute and chronic rodenticides.
- For effective and successful rodent control, the following programme should be adopted on large areas at a time on community approach.
- Chronic or multi-dose rodenticides (at present only anti-coagulants) are much safer than acute rodenticides because they are less toxic to the non-target species. But they have to be fed for 5 - 7 days to obtain desired results, which increases the cost of operation.

Day-1: Identify live burrows and place 15 g bromodiolone concentrate bait inside the burrow.

Day-2: Repeat Bromodiolone baiting in active or live burrows

Day-3: Eliminate residual population through trapping or fumigation with burrow fumigator

Principles

- Grow same maturity group cultivars on large areas to restrict the availability of the vulnerable stage (Reproductive) of the crop.
- Reduce the number and size of the bunds, keep them clean to locate burrows and avoid harborage.

- Rodent control operations should be taken up on large area at a time.
- It checks cross infestation or migration of rodents from untreated fields to treated fields.
- All the control operations should be completed before the crop attains primordial initiation stage since at this stage the rodents are invariably attracted to the rice crop.
- Rodenticides should be made available before beginning of the season.

References

Chapman, S.R. and L.P. Carter. 1976. Crop Production: Principles and Practices. San Francisco, USA: W.H. Freeman and Co. pp. 247-258.

FAOSTAT. 2014. Food and agricultural commodities production: commodities by regions. Retrieved Aug. 18, 2014 from http://faostat3.fao.org/browse/rankings/ commodities by regions/E.

Lantican, R.M. 2001. The Science and Practice of Crop Production. UPLB, College, Los Baños, Laguna: SEAMEO SEARCA and UPLB. pp. 4-5.

2

Biotic Stresses of Major Pulse Crops and Their Management Strategies

Reena and Surendra Prasad

India is the largest producer and consumer of pulses in the world, accounting for 33 per cent area and 25 percent production. Pulses are mostly grown under rainfed conditions in marginal to sub-marginal lands. Once a net exporter it is presently one of the largest importer of pulses as our demand outstrips domestic production. While the traditional cropping pattern almost always included a pulse crop either as a mixed crop or in rotation, the commercialization of agriculture has encouraged the practice of sole-cropping. As a result, the annual import has increased from 0.50 million tonnes to 1.80 million tonnes during the last five years and contribution of pulses in the national food basket reduced from 17% to 7%. Pulses are generally grown on marginal and sub-marginal lands under rainfed conditions with low inputs and suffer heavily due to biotic and abiotic stresses, resulting into low productivity. Challenges for pulses in India are decline in area of pulses, low genetic yield potential, low realized yield, instability in production, climate change, biotic and abiotic stresses, poor seed replacement rate, post harvest losses, wide fluctuation in price, no regular MSP/procurement policy, poor availability of critical inputs in productivity zone and poor transfer of technology.

Adoption of Integrated crop management module for pulses is very important. Biotechnological approaches for improving host plant resistance to insects is one of the most economic means of controlling insect pests. However, only low to moderate levels of resistance have been observed in the cultivated germplasm for *Helicoverpa* in chickpea. Diseases and pests are wide spread in pulses which include insects, fungal, viral, bacterial and nematodes. Chemical control for management is hazardous for human health as well as environment. Therefore, an integrated approach is required to control the problems. Integrated pest management involves the use of alternative techniques and options that are available and help keep the pest population below economic threshold (ET) level; this approach recommends use of chemicals as a last option for pest control.

More than 250 insect species are reported to affect pulses in India, among which, nearly one dozen cause heavy crop losses. On an average 2-2.4 million tonnes of pulses with a monetary value of nearly Rs 6,000 crore are lost annually due to ravages of insect pest complex (Reddy, 2009). Among them, pod borer (*Helicovera armigera)* causes the most harm, followed by wilt and root rot.

• Pulses may be classified as

– Kharif pulses

- Blackgram
- Greengram
- Rajmash / French bean
- Cowpea
- Cluster bean

-Rabi pulses

- Chickpea
- Field pea
- Lentil
- Lathyrus

Black gram / Green gram

It is inflicted by several major insect pests viz., Pod borer, hairy caterpillars, stem fly, whiteflies, thrips, aphids, pod bugs, etc.

1. Pod borer, *Maruca (testulalis) vitrata* (Geyer) (Lepidoptera: Pyralidae)

The greenish larvae with brown head, webs together the flowers and feeds on them and also bores into pods and feeds on the seeds causing extensive losses. Moth has dark brown forewings with a white cross band and white hind wings on a dark border.

2. Hairy caterpillar, *Amsacta moorei* (Butler)

***Amsacta albistriga* (Walker)**

Spilosoma obliqua Walker (Lepidoptera:* Arctiidae*)

Stoutly built moths appear with first shower, and lay yellow spherical eggs in clusters of 700 – 850 each on the under surface of leaves. The young caterpillars after hatching in 2-3 days, feeds on the photosynthetic part of the leaves, thus they become whitish in colour. Such leaves are easily recognizable from a distance and can be plucked along with the congregated larvae and destroyed. The later instars are voracious feeders and completely skeletonize the crop, in case of heavy infestation. The pest is active during rainy season (Mid –June till end of August) and passes the rest of the year in pupal stage in soil. Full grown larvae enter the soil and pupate at a depth of about 23 cm. They emerge next year at the onset of monsoon. This polyphagous insect feeds practically on all kinds of vegetation during kharif.

3. Stem fly *Ophiomyia phaseoli* (Tryon) (Diptera: Agromyzidae)

The maggots of metallic black flies bore into the stem thus causing withering and drying of the affected shoots, thus reducing the yield. Adults also puncture the leaves, which thereby turn yellow. Damage is more severe in seedlings than in grown up plants. The pest completes 8-9 generations in a year and passes winter as larva or as pupa.

4. Whitefly, *Bemisia tabaci* (Genn.) (Diptera: Aleyrodidae)

The nymphs are louse-like sluggish creature, while the adult are pale yellow, 1.0-1.5 mm long, with white waxy powdery covering. These small white coloured adults and nymphs suck the cell sap, thus devitalizing the plant and are also vector of Mungbean Yellow Mosaic Virus (MYMV) and Urdbean Yellow Mosaic Virus (UYMV). They also excrete honeydew on which the sooty mould grows, interfering with the normal photosynthesis process and giving the crop a sickly, black appearance from a distance.

5. Thrips, *Caliothrips indicus* (Bagn.) (Thysanoptera: Thripidae)

Small slender nymphs and adults damage the flowers and due to their damage the flowers shed before opening. Adult female lays kidney shaped eggs singly in the slits. Eggs hatch in 4-9 days and nymphs start feeding on plant juices by lacerating the leaf tissues. Infested inflorescence becomes abnormal showing symptoms of flowers drop. In case of severe incidence there is 100% loss in grain yield. Several generations are completed in a year.

6. Green Jassid, *Empoasca kerri* (Hemiptera: Cicadellidae)

These elongate (3 mm long), active and green adults and nymphs suck cell sap. The attacked leaves turn rust-red, show mottling, whitening and yellowing, cup downwards, dry up and fall to the ground. A single female lays 25 – 60 eggs in 25 – 30 days, which hatch in 4-11 days. Nymphal period is of 25 days, while the adult male and female longevity is 2 weeks and 13 weeks respectively. It breeds throughout the year.

7. Bean bugs, *Clavigralla gibbosa* Spinola (Hemiptera: Coreidae), Dusky cotton bug, *Oxycarenus laetus* Kirby (Hemiptera: lygaeidae)

Both nymphs and adults of the greenish brown bugs, suck cell sap from leaves, stem, pods, flower buds. Pods show yellow patches and later pods as well as grains shrivel in case of heavy attack. The grain size reduces thereby reducing the yield significantly. The pest is active from middle of October till end of May and has six overlapping generations during this period.

Greenish brown adult female bugs lay as many as 52 eggs eggs on pods. Both nymphs and adults suck cell sap from leaves, stem, pods and flower buds, forming yellow patches on pods. Later pods as well as grains shrivel in case of heavy attack, thus reducing the grain yield significantly. The mean egg, nymphal and adult periods of *C. gibbosa* on field bean is 8, 20 and 11 days. The total life cycle of *C. gibbosa* is 42 days and the pest is active from middle of October till the end of May and has six overlapping generations during this period.

O. laetus bugs lay transparent, light yellow, oval or cigar shaped eggs either singly or in groups of 2 - 10 usually on pods. Nymphs after hatching in 8 days, completes its nymphal period in 15 to 23 days. Adult bugs are small, flat with pointed head and uniform dark brown body with dirty white or dusky brown transparent hemielytra. Total life cycle varies from 35 to 50 days.

8. Bean mite, *Polyphagotarsonemus latus* (Banks) (Acarina: Tarsonemidae)

Small, elliptical, light, translucent yellowish green nymphs and adults suck cell sap from leaves, stem, pods, flower buds of both mungbean and urdbean. Reduction in photosynthesis and instability of water balance are some of the damaging effects to plants. As a result terminal leaves and flower buds become cupped and distorted. On the underside on the leaves corky brown areas are formed near the mid vein as a result of their feeding. Pods as well as grains shrivel in case of heavy attack.

9. Cut worm, *Agrotis ipsilon* (Hufnagel), *Agrotis flammatra* Schiff. (Lepidoptera: Noctuidae)

Larva become active at night and damages the seedlings by cutting at the base or just above soil surface. The cut plants or twigs are dragged and partially buried into the soil. This indicates their presence in the field. During day, the caterpillars remain hidden in cracks or within clods. Extremely dry weather during April – May adversely affects cutworm multiplication. During off-season (summer and rainy), the pest remains scattered among weeds growing nearby.

Management Strategies

- Early sowing during spring (15th March) suffers less damage protects the crop from peak incidence of whiteflies and thrips.
- Field sanitation, rogueing.
- Crop rotation.
- Deep summer ploughing and Pre-monsoon deep ploughing (two/three times) to expose the hibernating pupae to sunlight and predatory birds.
- Destruction of alternate host plants.
- Eradicate weeds and volunteer sesame plants.
- Seed treatment with phorate @ 1 kg per 50 kg seed gives protection for 2-3 weeks after germination.
- Inter/mixed cropping with taller crops like maize, sorghum, cotton, pearl millet, pigeonpea, etc., shields the crop, thereby checking the dispersal of flying insects.
- Removal and destruction of the affected branches during initial stages of attack.
- Collection and destruction of insect infested plant parts.
- Collection and destruction of eggs and early stage larvae. The leaves with young congregated larvae can be easily collected and destroyed.
- Handpick the older larvae during early stages of the crop.
- The infested shoots and pods may be collected and destroyed.
- Handpick the gregarious hairy caterpillars and the cocoons which are found on stem and destroy them in kerosene mixed water.
- Moths are strongly attracted to light. Setting up of light traps following the first shower of monsoon and continued throughout the period of emergence for about one month, greatly reduces pest incidence. Use light trap @ 1/acre and operate between 6 pm and 10 pm.
- Install pheromone traps @ 4-5/acre for monitoring adult moths activity (replace the lures with fresh lures after every 2-3 weeks).
- Erect bird perches @ 20/acre for encouraging predatory birds such as King crow, common mynah etc.
- Conservation of natural enemies through ecological engineering.
- Shaking the infested plants over vessels of oil and water or oily cloth gives very effective control of pod bugs.

- Egg masses of pod bugs should be collected and kept away from the cropped area as 30% of the egg masses are parasitized by *Gryon* sp. This practice will allow the parasite to be released in nature but will prevent the nymphs from causing damage.
- Braconids, *Microgaste*r sp., *Bracon kitcheneri* and *Fileanta ruficanda* parasitizes *Agrotis* larvae, while *Broscus punctatus* and *Liogrullus bimaculatus* predate on cutworms.
- Spray neem oil @ 5 ml/l as foliar sprays.
- Cutworms can be controlled by applying Lindane 2 per cent dust @ 12.5 kg/ha before sowing of Rabi crop.
- Spraying the crop at bud initiation stage with Dimethoate, Malathion, Metasystox or Phosphamidon effectively controls these pests.

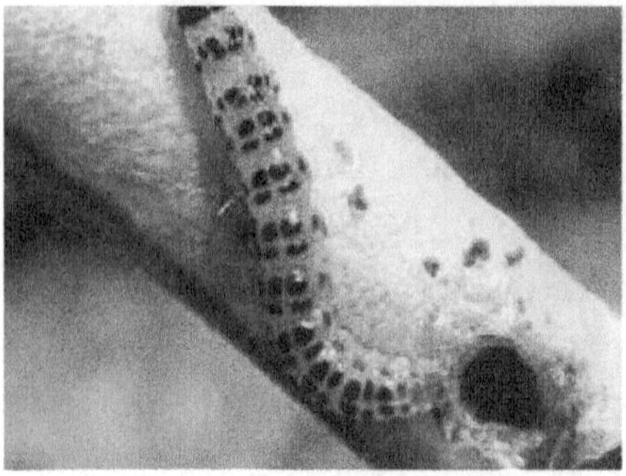

Maruca testulalis larva

M. testulalis adult

Hairy caterpillar damaged leaves

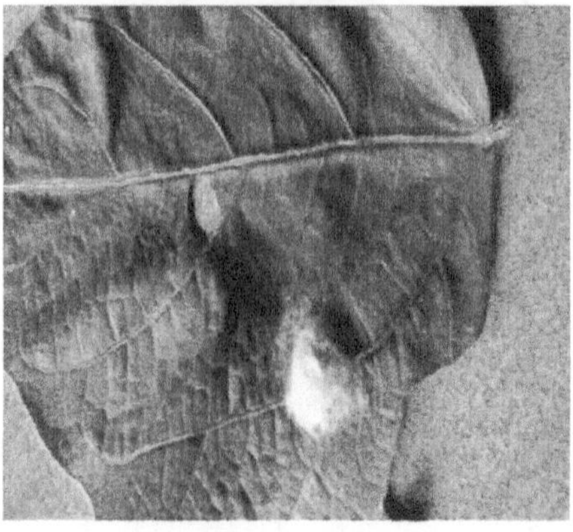

Hairy caterpillar egg masses

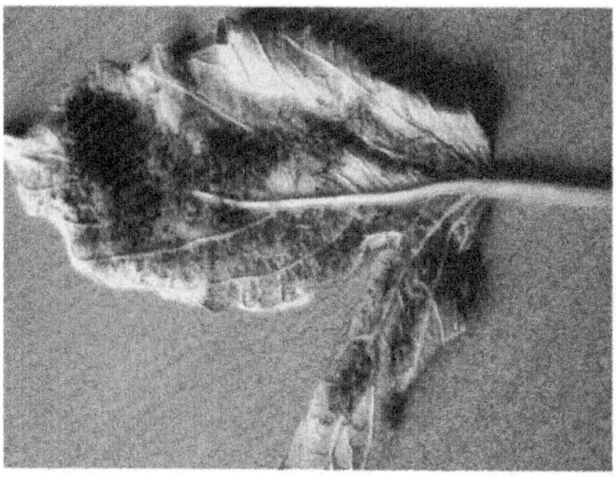

Young caterpillars after hatching

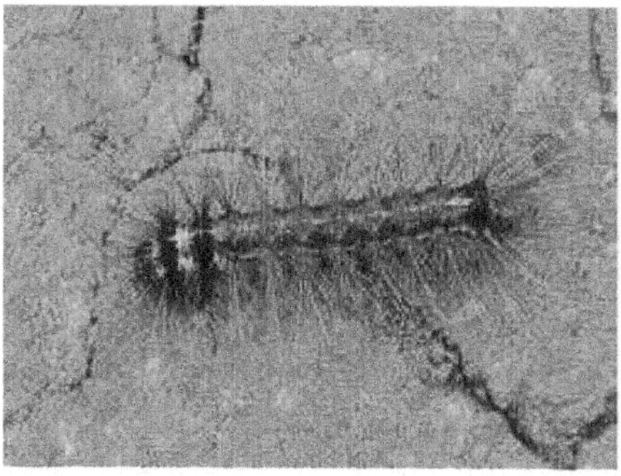

Fully grown caterpillar

Adult *Bemisia tabaci*

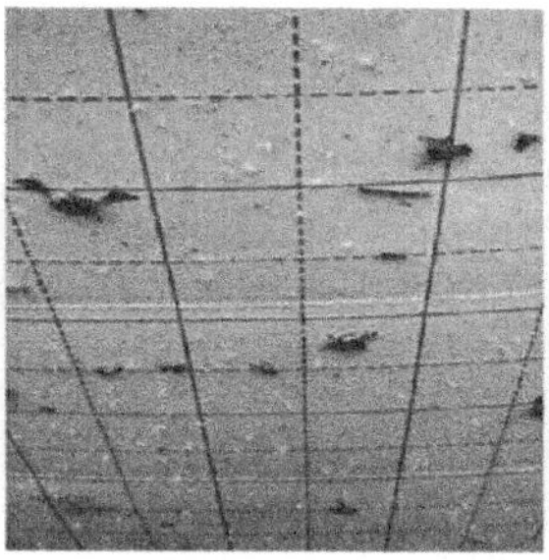

B. tabaci sticking to yellow sticky trap

Yellow mosaic affected urdbean

Caliothrips indicus adults

Empoasca kerri adult

Bean bug nymphs and adults damaging pods

Bean bug adult

Bean bug adult damaging pods

Polyphagotarsonemus latus adult

Cut worm larva

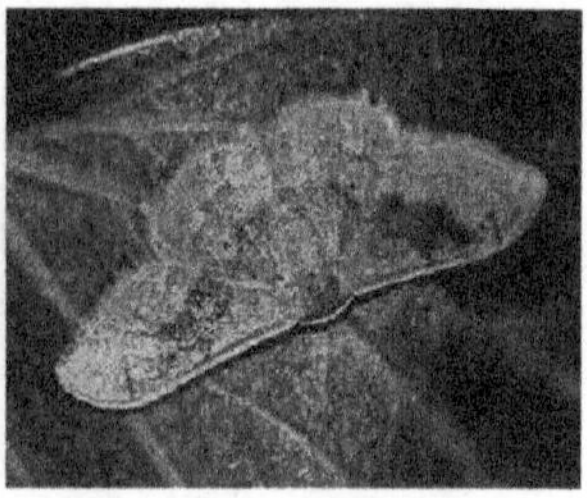

Cut worm adult

French Bean

1. Hairy caterpillar

- Same as for insect pests of urdbean.

2. Blister beetle, *Mylabris pustulata* (Thunberg) (Coleoptera: Meloidae)

- Same as in case of maize.

3. Bean fly *Ophiomyia phaseoli* (Tryon) (Diptera: Agromyzidae)

Minute, shiny black maggots mine sub-epidermally through the leaves. Later they tunnel the stem, resulting in drooping of the first two leaves and yellowing. Plants are most seriously affected at the seedling stage. Older plants also suffer leaf fall and stunting but are not usually killed. Females lay slender white eggs singly in holes made on upper leaf surface of young leaves. On hatching the larvae mine underneath the epidermis, until it reaches the midrib and then to the leaf stalk and stem. As many as seven generations are completed during the season.

4. Bean thrips, *Megalurothrips distalis* (Karny) (Thysanoptera: Thripidae)

Pale yellowish white nymphs and adults damages the flowers and due to their damage, flowers shed before opening. Infested inflorescence becomes abnormal showing symptoms of flowers drop. In case of severe incidence there is 100% loss in grain yield. Females are active flier and seek tender leaves on the host-plant for oviposition.

5. Cut worm

- Same as for insect pests of urdbean.

6. Bean mite

- Same as for insect pests of urdbean.

Management Strategies

- Timely sowing minimizes the attack.
- Seed treatment with Carbofuran @ 1 kg per 50 kg seed gives protection for 2-3 weeks after germination.
- Soil application of 10 kg Phorate 10G is effective upto 40 days of sowing.
- Foliar application of Dimethoate, Metasystox, Malathion, etc. may be given as per need.

Ophiomyia phaseoli adult

Megalurothrips distalis adult

Rabi Pulses

Chickpea

1. Pod borer, *Helicoverpa armigera* (Hubner) (Lepidoptera: Noctuidae)

The stoutly built, yellowish brown moths lay as amny as 741 greenish yellow round eggs on tender plant parts. Young larva after hatching in 2-4 days, feeds on tender leaves, buds, flowers and subsequently bores into pods and feeds on the seeds with its head and part of the body thrust inside, while the remaining part of the body remain outside. However, a change in feeding behavior has been noticed these days as reported by workers. The whole of *Helicoverpa* larva curls itself and hides inside the pod, while feeding on the seeds inside. A single larva may damage as many as 30-40 pods before undergoing pupation in soil.

2. Semilooper, *Autographa nigrisigna* Walker (Lepidoptera; Noctuidae)

Newly hatched larvae scratch chlorophyll from the leaves, while the grown up feed on leaves, buds, flower and pods by nibbling, leaving the basal part of pod with peduncle. The whole leaf becomes whitish and skeletonized as a result of feeding by scraping. Pod damage is similar to the damage caused by *H. armigera* that feeds on grains by making a neat hole on the pods. But this semilooper eats away everything leaving the peduncle only, very similar to the damage caused by birds. One generation is completed in 18-52 days. This pest is more active during early vegetative phase of the crop i.e. during November – December. With the advent of cold, its population reduces.

3. Aphids, *Aphis craccivora* Koch (Homoptera: Aphididae)

Oblong soft bodied, 2mm adults as well as nymphs suck the cell sap, due to which, there is depletion of assimilates coupled with increased respiration. The leaves and shoots are deformed and stunted followed by sticky honeydew deposition, in case of severe infestation. For early detection inspect new shoots and underside of young leaves. Apterous females (wingless) are shining dark brown or black, while the alate males (winged) are greenish black in colour. Both winged and wingless forms reproduce viviparously and parthenogenetically laying approximately 1743 numbers within 15 days. Breeding occurs throughout the year on one crop or the other. Besides chickpea, they attack several leguminous crops or weeds such as lentil, mungbean, lathyrus, fenugreek, alfalfa, tomato, potato, etc. Under dry conditions they multiply and spread over larger areas.

4. Hairy caterpillar, *Amsacta moorei* (Butler)

***Amsacta albistriga* (Walker)**

***Spilosoma obliqua* Walker (Lepidoptera: Arctiidae)**

5. Cut worm, *Agrotis ipsilon* (Hufnagel), *Agrotis flammatra* Schiff. (Lepidoptera: Noctuidae)

Management Strategies

- Field ploughing 4-5 times before sowing rabi crops breaks the clods in the field and reduces pest incidence.
- Clean cultivation, i.e. cleaning the weeds from bunds, nearby fields.
- Timely sowing i.e. upto mid October or growing early maturing varieties, so that it escapes the peak pod borer incidence.
- Wider spacing i.e 60 x 20 cm is more favorable to aphids than narrow spacing i.e. 30 cm.
- Close spacing or high plant population per unit area attract more pests. Therefore, close planting should be avoided.
- Cultivars influence population build up of aphids significantly.
- Deep ploughing of the fields in summer reduces the chances of pest propagation.

- Mixed or intercropping with non-preferred hosts like mustard, wheat, coriander, linseed, barley, increases the natural enemy activity and reduces pod borer incidence.
- Installation of pheromone traps @ 5/ha for monitoring and 20-30 /ha for mass trapping of the male moths.
- ETL (Economic threshold limit) is 5-6 moths of *H. armigera* / trap / day for few days during post winter months.
- Installation of 'T' shaped bird perches @ 50 /ha encourages feeding by insectivorous bird like, black drongo, common myna, etc.
- The leaves with young congregated larvae of hairy caterpillar can be easily collected and destroyed.
- Collection and destruction of grown-up hairy caterpillar larvae.
- Moths are strongly attracted to light. Setting up of light traps following the first shower of monsoon and continued throughout the period of emergence for about one month, greatly reduces pest incidence.
- Conserve larval parasitoid, *Campoletis chloridae* Uchida, as it may parasitize 50-60% of *H. armigera* larvae during vegetative phase and 30-40% during podding stage.
- A tachinid parasitoid has been recorded parasitizing larval and pupal stages of *A. nigrisigna.*
- NPV @ 250 - 500 larval equivalent, L.E. + adjuvants or provides effective control.
- Spray of NSKE (1 kg powdered neem kernel soaked in 20 litres of water over night + 200g of ordinary soap) @ 5%.
- *Beauveria bassiana* 1% WP, Azadirachtin 0.03% WSP or NPV of *H.a.* 2% AS may be sprayed.
- *Bacillus thuringiensis* formulations are also effective.
- Several coccinellids and syrphids predate on aphids.
- Braconids, *Microgaster* sp., *Bracon kitcheneri* and *Fileanta ruficanda* parasitizes *Agrotis* larvae, while *Broscus punctatus* and *Liogrullus bimaculatus* predate on cutworms.
- Carbaryl 10% DP, Chlorpyriphos 1.5% DP, Deltamethrin 2.8% EC, Novaluron 10% EC, Quinalphos 25% EC & 1.5% DP Phosalone (4% dust) @ 25 kg/ha or Malathion or dichlorvos spray gives effective control.

Gram pod borer larva, pupa and adult

Semilooper larva

Chickpea aphid

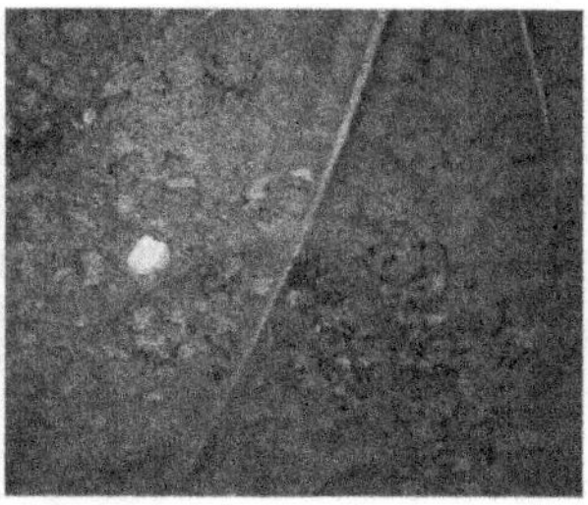

Hairy caterpillar egg mass First and second instar larvae *Amsacta albistriga*

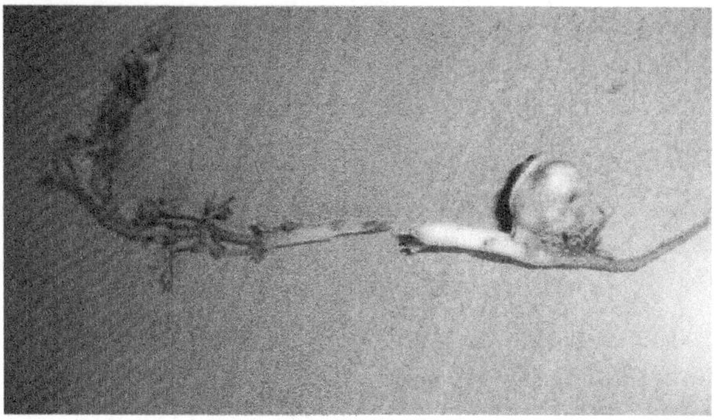

Damage caused by cut worm larva

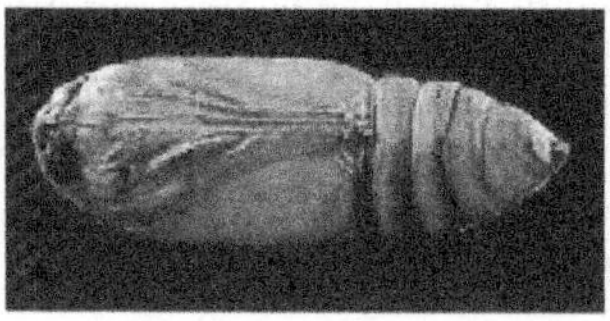

Cut worm larva, pupa and adult

Field pea

1. Pea leaf miner, *Chromatomyia horticola* (Goureau) (Diptera: Agromyzidae)

Greyish adult flies lay eggs singly in leaf tissues in the beginning of December. Larvae after hatching in 2-3 days damages by making serpentine, zig-zag mines in the leaves, usually from the periphery and ends towards the mid rib. As larva grows in 5 days, the mine widens and becomes distinct marked by dark excretal pellets. As a result of their feeding, photosynthetic area is reduced and growth of the plant, flowering and fruiting is hampered. They pupate inside the galleries itself. Adults emerge in 6 days and the total life cycle is complete in 13-14 days. Adults feed on cell sap with the help of their sucking and sponging type of mouth parts and also feed on pollen. There are several generations in a year.

2. Pea pod borer, *Etiella zinckenella* Treitschke (Lepidoptera: Phycitidae)

Grey moths lay eggs either singly or in clusters on various parts of plant preferably pods at reproductive stage. Newly emerged tiny greenish caterpillars after hatching in 5 days feed on floral parts and subsequently bore into the pods to feed on the seeds. Dropping of flowers and young pods is noticed as a result of their feeding, while the older pods are marked with a brown spot from where the larva enters. Up to 5% reduction in yield may be noticed as a result of their feeding. It completes 5 generations in a year and breeds throughout the year. It is a serious pest of lentil and green pea in northern India and it also attacks variety of other pulses in various parts of the country, Myanmar and Sri Lanka.

3. Gram pod borer

- Same as for insect pests of chickpea.

4. Pea stem fly, *Melanogromyza phaseoli* Tryon (Diptera: Agromyzidae)

Eggs are laid on the wall of the immature pod. Creamy maggots after hatching, damages by tunneling through the stem of young seedlings and later on the growing plants. Drooping of the tender leaves and wilting of seedlings is noticed when attacked at the early growing period. In case of severe infestation, leaves turn yellow giving a dry appearance to plants. Stem turn brown, becomes swollen and breakdown, where maggot and pupae are present. Attacked plants bear few pods, which are mostly empty or bear very small seeds.

5. Pea blue butterfly

***Lampides boeticus* Linnaeus**

***Euchrysops (Catochrysops) cnejus* (Lepidoptera: Lycanidae)**

Flat, slightly rounded, hairy larvae starts feeding on tender leaves. Later they bore into the buds, flowers, green pods and feed inside the pod on the developing grains.

6. Pea green aphid, *Acyrthosiphon pisum* (Harris) S. lat. (Hemiptera: Aphididae)

This oligophagous specie, feeds on only some species of legumes, while the other species, like green peach aphid, *Myzus persicae* (Sulzer) is a polyphagous specie, which feeds on over 400 plant species. Both adults and nymphs suck cell sap mostly from growing parts of the plants, thereby retarding development and sometimes causing complete loss of crop. Plants become stunted and yellow, while the leaves and pods are malformed and grains remain undersized.

7. Cut worm

- Same as for insect pests of chickpea.

Management Strategies

- Avoid early sowing. Sow the crop in 2nd fortnight of October to escape damage by this pest.
- Seed treatment with chlorpyriphos may be done before sowing.
- Spraying with Dimethoate 0.0%, when attack starts effectively checks the pest damage.
- Spraying may be repeated after 15 days, if necessary.
- Spraying with Carbaryl 0.05% at flower initiation stage effectively controls the pest for at least three weeks.

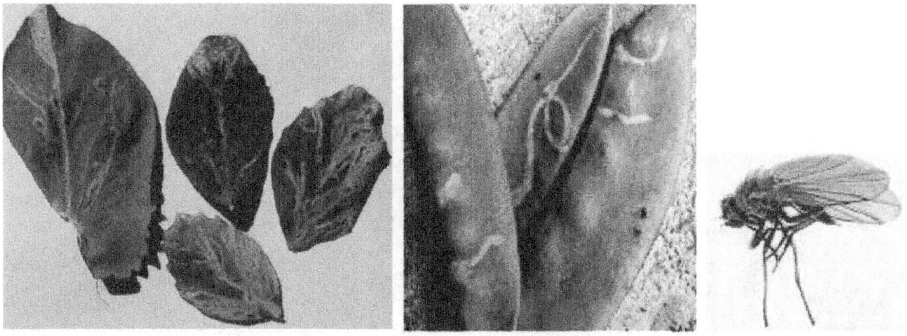

Pea leafminer larvae and pupae forming galleries on leaves and pods Miner adult

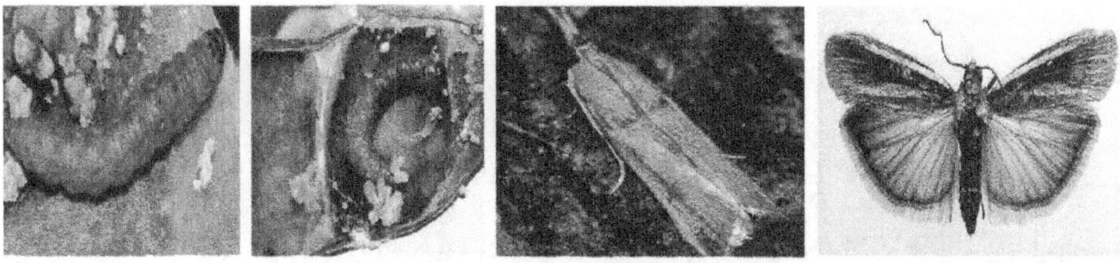

Pea pod borer larva and adult

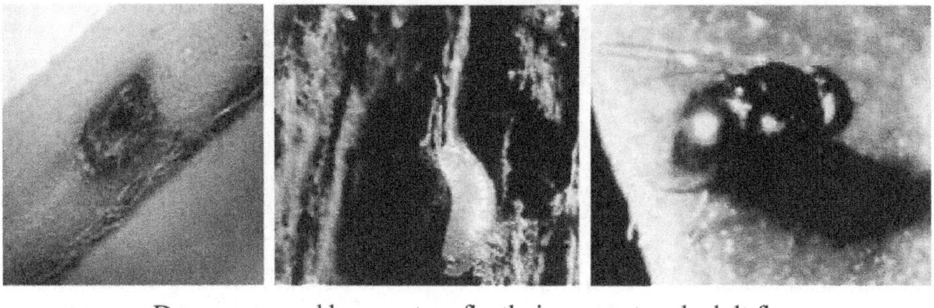

Damage caused by pea stem fly, their maggot and adult fly

Pea blue butterfly adult

Green pea aphid nymphs on leaves and pods

Pea aphid adult

Aphid parasitized by a parasitoid wasp

Lentil

1. Lentil pod borer, *Etiella zinckenella* (Lepidoptera: Phycitidae)

- Same as for insect pests of field pea.

2. Black aphids

- Same as for insect pests of chickpea.

3. Cut worm

- Same as for insect pests of chickpea.

Lathyrus

1. Pod borer, *Helicoverpa armigera*

- Same as for insect pests of chickpea.

2. Aphids

- Same as for insect pests of chickpea.

3. Bean bugs, *Clavigralla gibbosa* Spinola (Hemiptera: Coreidae), Dusky cotton bug, *Oxycarenus laetus* Kirby (Hemiptera: Lygaeidae)

- Same as for insect pests of urdbean / mungbean

Horse Gram

1. Pod borer, *E. zinckenella* (Lepidoptera: Phycitidae)

- Same as for insect pests of field pea.

2. Hairy caterpillar

- Same as for insect pests of chickpea.

3. Aphids, *Aphis craccivora*

- Same as for insect pests of chickpea.

4. Whitefly and Green Jassid

- Same as for insect pests of urdbean and mungbean.

5. Plume moth, *Exelastis atomosa* (Lepidoptera: Pterophoridae)

Larva is the damaging stage of this pest. Young larvae bore into the unopened flower buds for consuming the developing anthers. Grown up larvae bore into pods, however it never enters the pod completely.

Management strategies

- Spraying with Carbaryl 0.05% at flower initiation stage effectively controls the pest for at least three weeks.

Whitefly adults

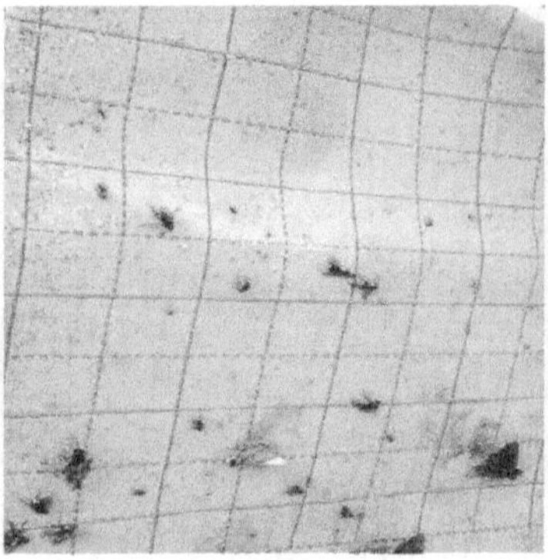

Yellow sticky trap

Plume moth adult

References

Atwal, A.S. and Dhaliwal, G.S. 1997. Agricultural Pests of South Asia and their Management. Kalyani Publishers. Ludhiana, India.

http://agritech.tnau.ac.in/crop_protection/crop_prot_crop_insect_agri_pest.html

http://www.pulseusa.com/docs/wanner2010.pdf

https://www.slideshare.net/SurabhiPal1/insect-pests-of-pulses-edited-23808

Srivastava, K.P. 1996. *A Textbook of Applied Entomology* (Vol. II). Kalyani Publishers, New Delhi, India.

3

Insect Pests of Oilseed Crops and Their Management

Manoj Kumar Jat, Arvind Singh Tetarwal and Ankit Kumar

The oilseeds crop occupies an important position in agriculture economy in the world. Indian agriculture has made considerable progress, particularly in respect of food crops such as wheat and rice in irrigated areas; however, performance has not been so good in case of other crops particularly oilseeds, pulses, and coarse cereals. Therefore, after achieving self sufficiency in food grains the government is focusing attention on these agricultural commodities. On the oilseeds map of the world, India occupies a prominent position, both in regard to acreage and production. India is the 4th largest edible oil economy in the world and contributes about 10 percent of the world oilseeds production, 6-7% of the global production of vegetable oil, and nearly 7 percent of protein meal. This sector also has an important place in the Indian agricultural sector covering an area of about 26.5 million hectares, with total production of over 25.3 million tonnes in the triennium ending 2015-16 (GOI, 2016). This constitutes about 14.8 per cent of the gross cropped area in the country. A wide range of oilseed crops is produced in different agro-climatic regions of the country. Insects-pests are one of the limiting factors in the production of oilseed crops. Management of these pest problems by using possible control techniques could increase the quality and quantity of the products. These crops are damaged by a more number of insect pests, of which some are more serious. These pests can be effectively controlled by the integration of different techniques such as use of various safe insecticides/biopesticides, some modification in cultural practices and use of pest tolerant varieties. Integrated pest management approaches will help to increase the production and productivity of oilseed crops by reducing the pest damage without any adverse effect on the agro- system and erosion in the environment.

Pest wise information of oilseed crop and management of major pests are described below.

Mustard / Rapeseed / Toria

This oilseed crop is inflicted by several insect pests of economic importance, viz., Mustard Aphid, Painted bug, Mustard saw fly, etc.

Mustard aphid, *Lipaphis erysimi* (Aphididae; Homoptera)

It is the most serious pest of the mustard crop in India. Besides brassicas to which mustard belongs, this pest attacks a number of other economic plants, particularly those of the family Cruciferae. Like many other important aphid pests, this species has a very wide distribution in the world. The mustard aphid is a most damaging pest of cruciferous oilseeds crops like Toria, Sarson, Raya, Taramira and *Brassica.* The aphids are minute, soft-bodied and light green insects having a pair of short tubes called cornicles on the fifth abdomen segment.

Life cycle: The insect reproduce partheno-genetically and the females give birth to average 26-133 nymphs. In favourable conditions they grow very fast and are full-fed in 7-10 days. Cloudy weather is most faourable for the multiplication of this insect. About 45 generations are completed in a year. The winged (alate) forms are produced in autumn and spring, and they move from one field to another field.

Fig. 1: Mustard plant attacked by aphid, *Lipaphis erysimi*

Nature of damage: Both nymphs and adults suck cell sap from leaves, stems, inflorescence and the developing pods of plants. Due to the very high population of the pest, the growth of plant greatly reduced. The leaves acquire a curly appearance, the flowers fail to form pods formation and not produce healthy seeds. The honeydew excreted by the aphids provides congenial conditions for the growth to development of sooty mould on the plant. In case of severe infestation the crop yield may be reduced up to 80 per cent.

Management Strategies

- Early sowing of mustard before 15th October will help to escape the attack of the pest and economic damage.
- Use tolerant varieties
- Three rounds of manual removal (clipping) of aphid infested twigs at 15 day intervals starting with the first appearance of the pest have been found effective if cheap labour is available.
- Installation of yellow sticky traps @ 10 traps/ ha to monitor the activity of aphids and to take the timely decision for foliar spray of insecticides before reaching the population at ETL level.
- Biological control: Ladybird beetles viz., *Coccinella septempunctata, Menochilus sexmaculata, Hippodamia variegata* and *Cheilomones vicina* are most efficient predators of the mustard aphid. A single adult beetle may feed an average of 10 to 15 adults/ day
- Botanical pesticides: Foliar spray of Neem oil 2 % and Neem Seed Kernel Extract (NSKE) 5 % effective against the mustard aphid
- Apply any one of the following insecticides when the population of the pest reaches 13-15 aphids per 10 cm terminal portion of the central shoot or when an average of 0.5-1.0 cm terminal portion of central shoot is covered by aphids or when plants infested aphids reach 20 per cent: Foliar sprays with 625 ml of oxydemeton methyl (Metasystox) 25 EC or 625 ml of dimethoate (Rogor) 30 EC or in 625 litres of water per ha or imidacloprid 0.01% or acetamiprid @ 0.01%.

Painted bug, *Bagrada hilaris* (Pentatomidae; Hemiptera)

The Painted bug *Bagrada cruciferarum* and *B. hilaris* (Burmester) have been recorded as major pests of various *Brassica* spp. as also other cruciferous crops and weeds. The painted bug is widely distributed in Myanmar, Sri Lanka, India, Iraq, Arabia and East Africa.

Life cycle: Painted bugs lay oval, pale-yellow eggs singly or in groups of 3-8 on leaves, stalks, pods and sometimes on the soil. A female bug may lay an average 37-102 egg in its life-span of 3-4 weeks. The eggs hatch in 3-5 days during summer and 20 days during December. The nymph moults and passing through five stages to attain adult stage in 16-22 days during the summer and 25-34 days during the winter. The entire life-cycle is completed in 19-54 days and it passes through 6-8 generations in a year.

Nature of damage: Both nymphs and adults suck cell sap from tender plant parts causing yellowing of leaves which gradually dry up and ultimately fall down exposing the plants to secondary invasion of bacteria and fungi. The plants wilt and wither affecting adversely the yield both quantitatively and qualitatively.

Management Strategies

- Deep summer ploughing to destroy eggs of painted bug
- Clean cultivation by removing weeds harbouring this pest is imperative for avoiding infestation of these bugs.
- Early sowing is beneficial to avoid pest attack.
- Seed treatment with Imidacloprid 70 WS or Thiamethoxam @ 5gm/ kg seed.
- The bugs usually congregate on the leaves and stem which can be jerked to dislodge them and killed in kerosin water
- Infected crop residues of mustard should be burned to avoid the painted bug infestation in next year crop.
- Biological control: Conserve bio-control agents such as Alophora spp. (tachinid fly) parasitizing eggs of painted bugs.
- Chemical control: In case of heavy infestation, spray with dichlorvos76% EC @ 250.8 ml in 200-400 l of water/acre or Imidacloprid 70% WS @ 700 g/100 Kg seed.

Mustard Sawfly, *Athalia lugens proxima* (Tenthredinidae; Hymenoptera)

Distribution

Mustard sawfly is a major pest of cole crops but it also affects almost all cruciferous plants, including rape and mustard. The peak period of activity is during September to December after which the activity declines; the pest is hardly noticed from March to July and appears on radish by the end of July.

Life cycle: A female lays on an average 35 eggs. Eggs hatched in 6 - 8 days. Newly hatched grubs are 2 -3 mm long, smooth, cylindrical and greenish-grey in colour; full grown ones are cylindrical in shape, 16 - 20 mm long and greenish-black in colour. They look and behave like caterpillars but have 8 pairs of prolegs. Grub development takes 21 - 31 days. Adults are 8 - 12 mm long, having dark head and thorax, orange coloured abdomen and translucent smoky wings with black veins. Females have a strong saw-like ovipositor – hence it has been given the popular name sawfly. They generally do not fly long distances but hop from leaf to leaf or fly from one plant to another plant. In South India where there is no severe winter, the pest undergoes as many as 10 overlapping generations in a year.

Nature of damage: Eggs are laid singly, mostly during day time and inserted into leaf tissues with the help of saw like ovipositor. On hatching the grubs nibble the margins of tender leaves but later on bite holes in the leaves. Grubs are diurnal in habit and feed generally during early morning and evening hours.

Management Strategies

- Deep summer ploughing of the field should be done to destroy the pupae.
- Apply irrigation in seedling stage for sawfly management because most of the larvae die due to drowning effect.
- Hand collection and destruction of larvae of saw fly during morning and evening hours.
- Release and conserve larval parasitoid, *Perilissus cingulator.*
- Chemical control: Seed treatment with imidacloprid 70% WS @ 700 g/ 100 Kg seeds. Foliar spray with dimethoate 30 % EC @ 264 ml or quinalphos 25 % EC @ 480 ml in 200-400 l of water/acre or spraying with malathion 50 EC @ 1000 ml/ha in 600-700 l water/ ha.
- Hand-picking of grubs which are not active during dawn and dusk if the area under crop is limited.

Sesame

Sesame (*Sesamum indicum* Lin.) is an important kharif oilseed crop of rainfed areas. It is known is one of oldest oilseed crops grown in India, known as 'queen of oil seeds'. India ranks first in area under cultivation representing 30% of the world production and Rajasthan, Maharashtra, Gujarat, Madhya Pradesh, Andhra Pradesh, Karnataka, Uttar Pradesh, West Bengal, Orissa, Punjab and Tamil Nadu are the major states of sesame cultivation (Singhal, 1999). It is attacked by several major insect pests viz. sesame leaf webber and capsule borer, *Antigastra catalaunalis*; gall fly, *Asphondylia sesame;* leaf hopper, *Orosius albicinctus*; sphingid moth, *Acherontia styx* and white fly, *Bemisia tabaci* etc. Out of these, sesame leaf webber and capsule borer is the notorious pest of sesame and causes up to 90% losses.

Leaf webber and capsule borer, *Antigastra catalaunalis* Duponchel (Pyralidae; Lepidoptera)

The sesame leaf and pod caterpillar is a serious and regular pest of Sesamum and is also distributed throughout India. The caterpillars are pale yellow, when young, but gradually become green and develop black dots all over the body. The full grown larva measures 14-17 mm. The moth is a small insect with a wing span of about 2 cm having dark brown markings on the wing-tips.

Life cycle: Females lying up to 140 eggs singly on the tender portions of plants at night. The eggs arc shiny, pale-green and they hatch in 2-7 days, depending upon the season. On emerging, the young larva, which measures about 2 mm in length, feeds for a little while on the leaf epidermis or within the leaf tissue. Soon after, it binds together the tender leaves of the growing shoot with the help of silken threads and continues to feed in the webbed mass. The size of this rolled mass increases gradually as the caterpillar grows older. It becomes full-grown in about 10 days in summer, but the period may be .prolonged to 33 days in winter. The grown-up larvae creep to the ground and pupate in silken cocoons in soil. Sometimes, pupation also takes place in the plant itself. Pupal development is completed in 4-20 days, depending upon the season. In summer, a generation is completed in about 23 days but in the winter it takes about 67 days.

Nature of damage: Young caterpillars feed on leaves of plants. They also bore into the shoots, flowers, buds and pods. An early attack kills the whole plant, but infestation of the shoots at a later stage hampers further growth and flowering.

Management Strategies

- Timely sowing of crop during June to July will escape from leaf webber damage.
- Collection and destruction of infested plants parts reduce the further damage of caterpillar.
- Conservation of existing predators like spiders, coccinellid beetles, stink bugs, preying mantids etc and parasitoids (Braconids and Ichneumonids) through application of botanical insecticides and safer chemicals.
- Spraying with Neem oil 2% or Neem Seed Kernel Extract (NSKE) 5% at the early stage of infestation would be effective and safer to natural enemies.
- Chemical control: Need based spray of carbaryl 50 WP 1000 g/ha in 500 litre of water

- Two sprays of 625 ml monocrotophos 36 SL or 500 ml diclorvos 76 EC or 1250 ml quinalphos 25 EC in 600-700 liters of water per hectare at pest appearance, flowering and pod formation or 30 and 45 days after sowing will control the pest effectively

Gall fly, *Asphondylia sesame* Felt (Cecidomyiidae; Diptera)

It is one important pest in south India and also in Rajasthan and a specific pest on gingelly.

Life cycle: The adult is a 5 mm long red-bodied midge. The small mosquito like fly inserts the eggs into the ovaries of flower buds. The pupation takes place inside the malformed capsule or pod. Life cycle is completed in 23 – 37 days. Activity of the pest starts at bud initiation and reach on its peak in September - November.

Nature of damage: Maggots feed inside the floral bud leading to formation of gall like structure which does not develop in to flower/capsules. The affected buds wither and drop. The larvae also feed on leaves and young shoots. Excreta (frass) remain between the leaves and the loose web. At a later stage, the larvae infest the sesame fruit capsule making an entrance hole on the lateral side and feeding on the seeds inside the capsule; they leave excreta on the seeds.

Management Strategies

- Natural enemies of gall fly: Conserve the existing parasitoids (*Pteromalus fasciatus*) and predators like spider, ladybird beetle, lacewing etc. to check the population of gall flies.
- Botanical pesticides: Foliar spray of Neem oil 2% or Neem seed kernels extract 5%, twice.
- Chemical control: Foliar spray of Quinalphos 25% EC @ 2000 ml in 500-700 liter of water/ha or Carbaryl 50 WP 1000 g/ha in 500 litre of water.

Leaf hopper: *Orosius albicinctus* Distant (Cicadellidae; Hemiptera)

The hoppers are light brown in colour and soft bodied insects. This is a national significance pest and distributed throughout the India. It multiplies very fast under favourable conditions.

Life cycle: On the basis of biological studies, mating occurs during dusk and last for 3-4 minutes. Both the sexes mate several times. A female lays many viable eggs. The eggs are inserted singly into midribs or veins, on the undersurface of petioles and stem of young plants. The average fecundity-cum-fertility is 6.5 and 140 during December-March and April-May, respectively.

The incubation period ranges from six days in June to 96 days in December; the nymphal period depends on the environmental conditions and being 11 to 107 days. Longevity of adults ranges from 19 to 105 depending on the season. The insect is predominant in summer.

Nature of damage: Both nymph and adult suck the sap from plants which leads to curling of leaf edges and leaves turn red or brown. In the later stages, leaves dry up and start shedding. The leafhopper also transmits sesamum phyllody disease.

Management Strategies

- Destroy all infested plant part to minimize pest damage
- Natural control of leaf hopper – Conserve the population of predators like spider, ladybird beetle, lacewing etc to maintain the population of this soft bodied insect below ETL.
- Chemical control: On the appearance on insect foliar spray of Oxydemeton–methyl 25% EC @1200 ml or Quinalphos 25% EC @ 2000 ml in 500-700 liter of water/ha.

Castor

Castor semilooper, *Achaea janata* (Noctuidae; Lepidoptera)

This is a serious pest of castor in all parts of India, Sri Lanka and Thailand. The adult of *A. janata* is a pale reddish brown moth with a wing expanse of 6-7 cm. The wings are decorated with broad zig-zag markings, a large pale area and dark brown patches. The full grown larva is dark and is marked with prominent blue-black stripes.

Life cycle: Female lays up to 450 eggs during its life span. The egg, being about 1 mm in length, is fairly large and also has on its surface a few ridges and furrows which radiate from the circular depression at the apex. The larva emerges by cutting a hole in the egg-shell in 3-5 days and devours it immediately. The larva feeds and moults 4-5 times and becomes full-grown in 15-20 days. The grown-up larva prepares a loose cocoon of coarse silk and some soil particles, and pupates under the fallen leaves on the soil, usually at the edge of the field. In some cases, pupation also takes place within the folded leaves on the plant itself. The pupal stage lasts 10-15 days and the moths, on emergence, feed on the soft fruits of citrus, mango, etc. There are 5-6 generations in a year.

Nature of damage: The caterpillars feed voraciously on castor leaves, starting from the edges inwards and leaving behind only the midribs and the stalks. With the excessive loss of foliage, the seed yield is reduced considerably.

Fig. 2: Affected leaves of castor plant by semi-looper, *Achaea janata*

Management Strategies

- Hand collection and destruction of the egg masses and first instar larvae.
- Spray of 0.05 % quinalphos 25 EC in 250 litres of water per acre and repeat at 15-day intervals.

Castor hairy caterpillar, *Euproctis lunata* (Lymantriidae; Lepidoptera)

The castor hairy caterpillar is widely distributed in India particularly in Uttar Pradesh, Orissa, Haryana, Madhya Pradesh, Andhra Pradesh, Karnataka and Tamil Nadu. It is observed feeding on linseed, groundnut and grapevine. Full-grown larvae are dark grey, with a wide white dorsal stripe, and have long hair all over body. The moths are pale yellow color.

Life cycle: The eggs are covered with the female anal tuft of brown hair. They hatch in 5-7 days and the young larvae feed gregariously for the first few days. Later on, they disperse and feed individually. They pass through six stages and are full-fed in 2-3 weeks. The full-grown caterpillars make loose, silken cocoons in the plant debris lying on the ground and pupate inside. The pupal stage lasts about one week in the summer. The pest passes through several generations in a year.

Nature of damage: Caterpillars feed on the leaves of various host plants and in case of severe infestation, they may cause complete defoliation. The attacked plants remain stunted and produce very little seed.

Management Strategies

- Deep summer ploughings to destroying the weeds and hibernating stages
- Use of light traps help in reducing the population of this pest.
- Hand collection and destruction of the egg masses and first instar larvae.

- Spray the crop with 200 ml dichlorvos (DDVP) 76 EC or 500 ml quinalphos 25 EC in 250 liters of water per acre.

Linseed

Linseed gall-midge, *Dasineura lini* (Cecidomyiidae; Diptera)

This insect appears as a serious pest of linseed in some parts of India, including Andhra Pradesh, Madhya Pradesh, Bihar, Uttar Pradesh, Delhi and Punjab. The adult is small delicate, mosquito like orange coloured insect.

Life cycle: The female lays 29-103 smooth, transparent eggs in the folds of 8-17 flowers or in tender green buds, either singly or in clusters of 3-5. The eggs hatch in 2-5 days. Just after emergence, the larvae are transparent, with a yellow patch on the abdomen. These larvae feed inside flower buds and eat the contents. They pass through four instars in 4-10 days and when full grown become deep pink and measure about 2 mm in length. The full-grown maggots drop to the ground, prepare a cocoon and pupate in the soil. The pupal period lasts 4-9 days. A generation is completed in 10-24 days. There are four overlapping generations during the season.

Nature of damage: Damage is the result of feeding by maggots on buds and flowers feed ovary. Consequently, no pod-formation takes place.

Management Strategies

- The adult killed by using light traps.
- The flies are also attracted in day-time to molasses or *gur* added to water.
- The incidence of this pest is more on the late-sown crop as compared with the normal-sown crop, the practice of normal-sown crops should be adopted if possible.
- If pest incidence is more (10 per cent) then spray the crop with 600 gm carbaryl 50 WP in 200 liters of water per acre.

Sunflower

Sunflower (*Helianthus annus* L.) is an important oilseed crop in India. Among oilseeds, sunflower commonly known as 'Surajmukhi' is one of the potential oil yielding crops gaining popularity because of its wider adaptability to different agro-climatic conditions. The production of this crop is seriously affected by the insect pests, attacking at different stages of crop growth. These losses can be minimized by adopting effective pest management strategies. Few major pests and their management practices have been discussed below.

Head borer, *Helicoverpa armigera* (Noctuidae; Lepidoptera)

The head borer or capitulum borer, *H. armigera* is highly polyphagous with about 181 host plants including important crop plants such as pulses, cotton, vegetables, oilseeds etc. and the pest is cosmopolitan in nature (prevalent throughout India).

Life history: The life cycle of this pest depends on the climatic conditions from north to south; it completes four generations in Punjab while seven to eight generations in Andhra Pradesh. Emergence of *H. armigera* moth has been observed evening any time after 04:00 p.m. The peak emergence being between 08.00 p.m. and 10.00 p.m. Female moth lays average 700-1000 eggs. The incubation period ranges from 2-5 days. There are normally six instars, but exceptionally seven instars are found in cold season. The larval period ranges from 8 to 33.6 days with 8 to 12 days on tomato, 21-28 days on chickpea, 21 - 28 days on maize, 33.6 days on sunflower and 20-21 days on cotton. The full grown larvae pupate in earthen cocoons in the soil. Pupal period vary from 5 - 8 days in India.

Nature of damage: The head or capitulum borer causes considerable damage to developing grains in the head capsule. The young larvae first attack the tender parts like bracts and petals, and later on shift to reproductive parts of the flower heads. Bigger larvae mostly feed on seeds by making tunnels in the body of the flower heads and often remain concealed. They may also shift to the backside of the heads and even leaves, and feeding may continue upto maturity. Star bud stage of the crop is most vulnerable and suffers maximum yield loss.

Management Strategies

- Early sown crop usually suffers lower attack of the pest.
- A significant reduction in pest density is achieved with the spray of HaNPV @250-500 LE/ha.
- Neem based insecticidal formulations such as Neem oil 2% or NSKE 5% are found effective in reducing damage due to *H. armigera*.
- In case of severe attack especially on late sown crop, when one larva per plant, average of 20 plants is present, spray the crop with 1500 ml of quinalphos 25 EC in 500 liters of water per hectare at the initiation of star bud Stage, and repeat after two weeks if necessary.

Bihar hairy caterpillar, *Spilosoma obliqua* (Arctiidae: Lepidoptera)

Hairy caterpillars are highly polyphagous pest found throughout the year. Among various hairy caterpillars, Bihar hairy caterpillar is major ones causing severe damage to the sunflower crop. Besides sunflower, it infests millets, cotton, jute, sunhemp, castor, cauliflower, cabbage etc. It has been reported to feed on 96 plant species in India. They are called hairy caterpillar because they have profused hairy growth on their body in larval stage.

Life cycle: The female lays eggs in cluster on the lower surface of leaves. After hatching, the tiny larvae feed gregariously on the chlorophyll content of the leaf upto second instar. The larva defoliates the plants and move from one field to another. The full grown larva is darkened with yellowish brown abdomen having numerous pale white brown and black hairs and measures about 43 mm. It pupates in soil. The adult is dull yellow with oblique line of black dots on hind wings. The dorsal side of the abdomen is red with dull yellow ventral side.

Nature of damage: The attacked leaves look like a dirty paper, which can be recognized from a distance. After this stage larvae start dispersing throughout the field and feed voraciously leaving only the veins of the leaves without any green material. The full grown larvae are more harmful. After finishing the foliage of one field they migrate to the adjacent field resulting in complete destruction of the crop.

Management Strategies

- Use of well rotten manures.
- Intercropping with pigeon pea at a row ratio of 2:1 is effective in reducing the insect attack. Hand collection and destruction of egg masses and gregarious larva should be done. The leaves on which large numbers of first instar larvae feed gregariously can also be collected and destroyed mechanically.
- To monitor the adult moths light trap should be installed in the field and attracted moths should be destroyed.
- Biological control: Application of *Bacillus thuringiensis* (Bt) @ 1.0 Kg/ ha has been found effective in controlling hairy caterpillars.
- Botanical insecticides: Spray 5% neem seed kernel extract preferably in the evening.
- Chemical control: Need based application of cypermethrin 10% EC @ 650-700 ml/ha diluted in 500 – 650 liter of water or quinalphos 25 EC @ 1.5 ml/ liter of water should be done in the evening. Spot application of chlorpyriphos 20 EC 1.0 ml/litre of water is highly effective for the control of gregarious phase larvae.

- Digging trench around the field and dusting them with carbaryl 10% dust prevents the migration of caterpillars from one field to another. 7.

Jassid, *Amrasca biguttula biguttula* Ishida (Cicadellidae: Homoptera)

This pest is of economic importance in Maharashtra, Tamil Nadu and Karnataka causing crop loss up to 46 %. Though it may appear on the crop round the year, it is serious during certain months at different places. Summer crops are likely to suffer more with this pest than kharif crop.

Life history: The female lays on an average 15 eggs into the spongy parenehymatous tissue between the vascular bundles and the epidermis and they hatch in 4-11 days. The nymphs moult five times and the whole life cycle is completed in two weeks to one and half month depending upon the temperature and humidity prevailing in the field.

Nature of damage: The incidence would start from seedling stage and prevail right through entire plant life. Stunted growth of plant, cupped and crinkled leaves, burnt appearance of leaf margins are symptoms of damage.

Management Strategies

- Seed treatment with imidacloprid @ 5 ml/kg of sunflower seed protects from sucking insects up to 35-40 days after sowing.
- Need based foliar spray with any one insecticide like phosphamidon (0.03%) or dimethoate (0.03%) or imidacloprid (0.01%) may be mixed in 650-700 litre water per hectare.

Groundnut

Ground aphid, *Aphis craccivora* (Aphididae; Homoptera)

Aphid is the most serious pests of groundnut in India. It also attacks peas, beans, pulses, safflower and some weeds. Its distribution is throughout India. It has also been recorded in Africa, Argentina and Chile. The winged adults are soft-bodied insects with black wings and they reach the freshly germinated groundnut plants after over-wintering on collateral host plants.

Life cycle: The females are parthenogenesis may produce 8-20 young ones in a life span of 10-12 days. The young nymphs are brownish and they pass through four moults to become adults in 5-8 days. The apterous females start producing brood within 24 hours of attaining that stage.

Nature of damage: The nymphs and adults suck the cell sap, usually from the underside of leaves. Infestation in the early stages causes stunting of the plants as well as reducing their vigour. When the attack occurs at the time of flowering and pod formation, the yield is reduced considerably. Infestation on the groundnut crop usually occurs 4-6 weeks after sowing. The aphid causes rossette disease of groundnut.

Management Strategies

- The pest appears on growing points, spray 500 ml of malathion 50 EC or 500 ml dimethote 35 EC in 500-600 litres of water per hectare.
- Detail of management practices see in mustard.

White grub, *Holotrichia consanguinea* (Melolothidae; coleoptera)

Whitegrub, *Holotrichia consanguinea* is the most serious scarab pest in India. This is a polyphagous pest in nature and prefers light sandy soils. This species predominantly found in Rajasthan, Gujarat, M.P., U.P., Haryana, Punjab, Bihar. Rajasthan and Gujarat has a long history of whitegrub, *H. consanguinea* infestation in all most the Kharif crops. The damage in different crops ranges from 20 to 100 per cent. This pest most preferably feed on groundnut and bajara.

Life cycle: The biology, of *H. consanguinea* has been worked out in Gujarat State by Patel *et al.* (1967), whereas in Rajasthan (Rai, *et al.,* 1969). Adult beetles lay eggs singly up to a depth of 9-10 cm. The eggs hatch in 8-10 days. The newly hatched grubs measure about 12 mm in length and their development is completed in 8-10 weeks. After the rainy season full-grown larvae migrate in soil to the depth of 40 to 70 cm or more in search of suitable moisture zone for pupation. The pupa is semicircular and creamy white and the pupal stage lasts about a fortnight. The beetles remain in the soil at a depth of 10-20 cm and come out for feeding at night. Adults formed in November remain in soil till next June. The beetles remain in the soil in inactive state upto middle of May at a depth of about one meter. The average duration of one life cycle is 122 days and there is only one generation in a year.

Nature of damage: The beetles of *H. consanguinea* emerge from the soil during dusk after good pre monsoon or monsoon rain in mid May. The beetles are polyphagous, and may feed on the foliage of a wide variety of host trees and bushes found in the nearby places. However, they have some preference for certain hosts like Jujube (ber), Khejri, *Prosopis cineraria*, Neem, *Azadirachta indica*, Cluster fig (gular), Jambolana (Jamun) and Drumstick (sanjana).

The grubs makes chamber by compressing the surrounding soil particles and then eats the rootlets exposed into the chamber; thereafter it little bit moves vertically to eat more of the same root. Then the, grubs move horizontally making chambers and feeding on the exposed roots. The grubs continue active feeding from July to mid-October.

Management Strategies

- The cultural management of white grub through deep summer ploughing exposes the grubs which are fed by birds and adult of grub can be killed by light trap.
- Annihilation of whitegrub beetles on host trees by application of insecticides and pheromone loading of selected host trees.
- Among the microbial control agents, entomopathogenic fungi, *Metarhizium anisopliae* and *Beauvaria bassiana* were more effective, when placed at soil depth of 10-15 cm. Nematodes, *Steinernema glaseri, Heterorhabditis sp.* and a local strain were found to be pathogenic to several whitegrub species including *H. consanguinea.*
- Seed furrow application of Phorate 10G @ 20 kg/ ha or Quinalphos 5G @ 30 kg/ha or Thiamethoxam 70 WP @ 80 a.i. /ha in pearl millet crop.
- In groundnut, seed treatment with Clothianidin 50 WG (Dantotsu) @ 2g/ kg seed or Imidacloprid 17.8 SL @ 3 ml/ kg seed.
- In standing crop, chlorpyriphos 20 EC or quinalphos 25 EC @ 4l/ ha applied with irrigation water after 3 weeks of first shower of monsoon.

Red hairy caterpillars, *Amsacta albistriga* Wlk. and *A. moorei* Butler (Arctiidae: Lepidoptera)

This is a serious and devastating pest of rainfed groundnut crop. It is an endemic pest and its seasonal outbreak in various areas is largely dependent on the climatic conditions and the local agricultural practices of the areas.

Life history: The adults are medium sized moths. After the receipt of heavy rains on the second evening at about 4 p.m. the moths emerge from their earthen cells in the soil. The moths mate and commence oviposition on the same day. The egg laying may last for 2 - 6 days. The creamy or bright yellow eggs are laid in groups mostly on the under surface of cowpea leaves usually sown along with groundnut as an inter crop and also on groundnut and occasionally on other vegetation, clods of earth, stones, dry twigs, etc. A female moth lays about 600 - 700 eggs but it has also been observed that as many as 2300 eggs have been laid by a moth. The incubation period ranges from 2 - 3 days. The larva becomes full grown in 40 - 50 days. It is about 5 cm long, reddish brown with hairs all over the body arising on warts. With the receipt of some showers, the grown up larvae burrow into the moist soil and pupate in earthen cells at a depth of 10 - 20 cm.

Nature of damage: Newly hatched larvae feed gregariously by scarping the under surface of tender leaflets leaving the upper epidermal layer intact. As they grow they feed voraciously on leaves leaving behind the petiole and midribs of leaves and the main stem of plants. They may be seen marching from one field to another in thousands. Severely damaged crop fields present the appearance as though the entire area has been grazed by cattle. Often it results in total loss of pods.

Management strategies

- Deep summer ploughing to destroy the pupae of this insect.
- Setting bonfires or light traps to monitor and attract the adult moths at night.
- Collection and destruction of egg masses should be carried out during the early stages of attack.
- After the mass emergence of moths, the field should be dusted with phosalone 4% or carbaryl 10% dust to kill the first instar larvae which are vulnerable at this stage.
- Grown up larvae are killed by spray application of phosalone 0.05 %.
- Nuclear polyhedrosis virus @ 250 LE/ha has been found promising in field scale control of the pest in Tamil Nadu.

References

Bakhetia, D.R.C. and Sekhon, B.S. 1989. Insect-pests and their management in rapeseed mustard. *J. Oilseeds Research*, 6 (2): 269-299.

Chander, S. and Phadke, K.G. 1994. Economic injury levels of rapeseed aphid, *Lipaphis erysimi* determined on natural infestation and after different insecticides treatments. *International Journal of Pest Manag*ement, 40:107-110.

Kalra, V.K., Gupta, D.S. and Yadav, T.P. 1983. Effect of cultural practices and aphid infestation on seed yield and its component taits in *Brassica juncea* (L.) Czern and Coss. *Haryana Agricultural University Journal Research.,* 13:115-120.

Nain, Rohit, Dashad, S.S. and Singh, S.P. 2009. Relative efficacy of newer insecticides against pod borer, *Helicoverpa armigera* (Hubner) infesting sunflower crop. Proc. National Symposium on role of pesticide application technology in crop protection: towards sustainability in agriculture.20-22 January, 2008 organised by Institute of Pesticide Formulation Technology,Gurgaon, India.pp.61-62

Patel, R. M., Patel, G. G. and Vyas, N. 1967. Further observations on the biology and control of white grubs (Holotrichia sp. near consanguinea Bl.) in soil affecting groundnut in Gujarat. *Indian J. of entomology*, 29(2): 170-176.

Rai, B. K., Joshi, H. C., Rathore, Y. K., Dutta, S. M. and Shinde, V. K. R. 1969. Studies on the bionomics and control of white grub Holotrichia consanguinea Blanchin lalsot, distt. Jaipur, Rajasthan. *Indian J. of Entomology*, 31(2): 132-142.

Singh, S.P. 2009. Insect pest management in oilseed crops. *Indian Farming*. 58(7): 29-33.

Singhal V. 1999. Indian Agriculture. Indian Council of Agricultural Research, New Delhi.pp. 600.

4

Biotic Stresses of Vegetable Crops and Management

Amandeep Kaur and Smriti

Vegetable is an important constituent of Indian diet. India is the second largest producer of vegetables which contributes 16.7% to global vegetable area with a production of 156.33 million tons (15.4%) (Table 1). The level of productivity of vegetable17.4 tons/ha is quite low (Rai *et al*., 2014). Majority of Indians are vegetarian, with a per capita consumption 135 g per day as against the recommended 300 g per day. It is still very less than recommended diet level. In near future, there is a need of around 5-6 million tons of food to feed our 1.3 billion Indian population expected by the year 2020 (Paroda 1999). The major biotic stress in vegetable production is due to increased insect pest. In many cases, there is 100 per cent yield loss due to viral diseases vectored by insects. Vegetables are more prone to insect pests and diseases mainly due to their tenderness and softness as compared to other crops and virtual absence of resistance characters because of intensive hybrid cultivation. The insect pests inflict crop losses to the tune of 10-30 per cent in vegetable production (Table 2). To overcome these losses farmers used indiscriminate doses of toxic chemical on vegetable. It accounts for 13-14 per cent of total pesticides consumption, as against 2.6 per cent of cropped area (Sardana, 2005). For the benefit of the farmer and other, application of BIPM is must. Bio intensive pest management involves every part of management that keeps pest below economic injury level so that pest does not cause any economic loss to the farmers or growers.

Table 1: Leading vegetable producing states of India

Sl. No.	State	State Area (m ha)	Production (mt)	Productivity (t/ha)
1.	Odisha	0.87	8.70	10.00
2.	Uttar Pradesh	0.86	13.88	16.13
3.	Bihar	0.85	12.28	14.44
4.	West Bengal	0.51	5.39	10.56
5.	Karnataka	0.29	5.70	19.65
6.	Kerala	0.24	2.78	11.58
7.	Maharashtra	0.21	2.95	14.04
8.	Tamil Nadu	0.17	4.39	25.82

(Dhandapani *et al.*, 2003)

Table2: Percent yield loss in vegetable in India due to insect pests

Crop	Insect pest	% yield loss
Cabbage and cauliflower	Diamond back moth (*Plutella xylostella*)	17-99
	Caterpillar (*Pieris brassicae*)	69
Brinjal	Fruit and shoot borer (*Leucinodes orbonalis*)	11-93
	Root knot nematode (*Meloidogyne* spp	27.2
Okra	Shoot and fruit borer(*Earias vittella*)	23-54
	Leafhopper (*Amrasca biguttula biguttula*	54-66
	Whitefly (*Bemisia tabaci*)	54
	Fruit borer (*H. armigera*)	22
Bitter gourd	Fruit fly (*Bactrocera cucurbitae*)	60-80
Cucumber	Fruit fly (*Bactrocera cucurbitae*)	20-39
Musk melon	Fruit fly (*Bactrocera cucurbitae*)	76-100
Snake gourd	Fruit fly (*Bactrocera cucurbitae*)	63
Tomato	Fruit borer (*Helicoverpa armigera*)	24-65
Chilli	Thrips (*Scirtothrips dorsalis*)	12-90
	Mites (*Polyphagotarsonemus latus*)	34

(Rai *et al.*, 2014)

Insect Pests attacking the major group of vegetable crops

1. Cole vegetables

1.1. Cabbage butterfly

1.2. Diamond back moth

1.3. Head borer

1.4. Tobacco caterpillar

1.5. Aphids

2. Brinjal

2.1. Fruit-shoot borer

2.2. Hadda beetle

2.3. Red spider mite

2.4. Whitefly

3. Okra

3.1. Spotted bollworm

3.2. Jassid

3.3. Whitefly

4. Cucurbits

4.1. Fruit fly

4.2. Red pumpkin beetle

5. Peas

5.1. Pea leaf miner

5.2. Pea aphid

5.3. Pod borer

6. Potato

6.1 Potato tuber moth

6.2 Cut worms

7. Tomato

7.1. Fruit borer

7.2. Whitefly

7.3. Mealy bug

7.4. Nematode

8. Chilli

8.1. Thrips

8.2. Whitefly

8.3. Mite

8.4. Fruit borer

Important approaches for pest management

Indiscriminate use of chemical pesticides has resulted in emergence of more aggressive pests due to resistance development; residual problems in food and drinking water and ecological imbalance due to elimination of beneficial microorganisms and insects. Therefore, for sustainability of vegetable crops, these biotic stresses need to be managed through eco-friendly measures supported by need based and judicious use of chemicals to achieve high economic returns without disturbing environmental balance and serenity. Integrated pest management (IPM) is one of the economically viable and environmentally safe key technologies to increase vegetable productivity in the country. Some of the important components of IPM practices in vegetables are given below:

- Use of resistant varieties/hybrids/genotypes (host plant resistance).
- Use of healthy seeds obtained from a reliable source.
- Crop rotations, intercropping, trap/barrier crops *etc.*
- Optimum dates for planting and harvesting.
- Use of biocontrol agents, biofumigants, botanicals *etc.*
- Need based application of safer and label claim chemical pesticides.

Table 3: Economic Thresholds (ETL) level for major vegetable pests.

Crop	Insect pest	ETL (s)
Cabbage and Cauliflower	Diamondback moth (*Plutella xylostella*)	2 larvae plant at 1-4 weeks after transplanting or 5 larvae plant at 5-10 weeks after transplanting or 1-5 % incidence
	Tobacco caterpillar (*Spodoptera litura*)	1-5 % incidence
Okra	Okra Jassids (*Amrasca devastance*)	4.46 nymphs per plant
	Okra fruit borer (*Earias vitella*)	5.3% infestation of fruits
Brinjal	Fruit shoot borer (*Leucinodes orbonalis*)	1-5% shoot/fruit infestation
Peas	Pea aphids (*Acyrthosiphon pisum*)	3-4 aphids per stem tip
Potato	Cut worms (*Agrotis ipsilon*)	One larvae per ten plants
Tomato	Tomato fruit borer (*Helicoverpa armigera*)	8 eggs per 15 plants or one larva per plant or 1-5 % incidence
	Leafminer	2-5 miners per plant
Chilli	Chilli mites (*Polyphagotarsonemus latus*)	One mite per leaf
	Chilli thrips (*Thrips tabaci*)	2 thrips per leaf
Radish	Radish aphids (*Acyrthosiphon pisum*)	75 aphids per plant

(Rai *et al.*, 2014 and Dhandapani *et. al.*, 2003)

Insect pests, diseases and nematodes are major constraints in achieving full yield potential in vegetables. The losses due to these biotic stresses are around 40 per cent and if we add about 15-20 per cent post-harvest losses, the situation becomes more alarming. In fact a typical eco-friendly approach for sustainable management of vegetable pests involves biological control including deployment of host resistance, best cultural practices and need based use of chemical pesticides. Such plant protection measures are also in harmony with international food safety and environmental protection protocols.

Prevention: Use practices that contribute to crop protection for the long term. These include: Biological controls. Crop rotation; breaks pest life cycles, often improves tilth and fertility. Host plant resistance; Use varieties that are resistant to common pest species. Sanitation; Remove or destroy debris and other sources of pest infestation. Site selection; Plant only on sites suited to the crop needs Collect valuable information in time to use it in making good decisions. Which of the expected pests are in your field? Know both "what" and "how many" by properly sampling the field. Use recommended guide techniques to accurately and efficiently collect this information

Management Strategies

For control measures choose those that optimize cost and effect while minimizing adverse effects. These are:

Cultural: eg. Crop rotation

Mechanical e.g. Destruction of pests, infested plant

Biological eg. Biological control agents

Genetic eg. Plant pest/disease-resistant varieties

Chemical eg. Insecticides, fungicides

Biological control: Study and utilization of natural enemies of insect like predators, parasites and pathogens by man to manage pest population below economic injury level is called biological control of insect pest.

Conclusions

Chemical measures are the most common method of pest management. Hence their judicious use is advocated which includes avoiding prophylactic sprays, adopting strip treatment, spot application to only those areas with heavy incidence of pests, applying to the soil to avoid direct contact with natural enemies and using selective or non-persistent pesticides. In vegetables skip row treatment with pesticides is given. Safer pesticides have been identified for use in conjunction with natural enemies.

References

Dhandapani, N., Umesh Chandra, R. Shelkar and M. Murugan. 2003. Bio-intensive pest management (BIPM) in major vegetable crops: An Indian perspective. *Food, Agriculture & Environment* Vol.1 (2): 333-339.

Paroda, R.S. 1999. For a Food Secure Future. The Hindu Survey of Indian Agriculture.

Shivalingswami, T.M., Satpathy, S. and Banergee, M.K. 2002. Estimation of crops losses due to Insect pests in vegetables Ed. by Sarath Babu, B.,Varaprasad, K.S. Anitha, K., Prasada Rao, R.D.V.J., Chakrabarthy S.K. and Chandurkar, P.S. Resource management in plant protection.1: 24-31.

Srinivasan, K. 1993. Pests of vegetable crops and their control. In: Advances in Horticulture, 6: Vegetable crops Eds. K.L. Chadha and G. Kalloo. Malhotra Pub. House, New Delhi. p. 859-886.

Sardana, H.R. 2005. Integrated pest management in vegetables. In: Training manual -2, Training on IPM for Zonal Agricultural Research Stations. pp. 105-118

5

Insect Pests Infesting Major Vegetable Crops and Their Management Strategies - I

Amandeep Kaur, R. M. Srivastava, S. K. Maurya and Tanuja Phartiyal

Vegetables constitute a substantial part of human diet supplying vitamins and minerals, in which other food materials are deficient. India has emerged out as the second largest producer of vegetables in the world. More than 25% of fruits and vegetables are damaged by the attack of insect-pests (Pradhan, 1964).Vegetable crop is damaged by both sucking as well as chewing insect-pests. Sucking pests are aphids, whitefly, jassids and bugs attacking in initial stage whereas chewing pests like borers, infesting in flowering and fruiting stage. Sucking pests may transmit the viral diseases which cause severe loss in yield in tomato and okra crops. About 13-14 % of total pesticides used in the country are applied on vegetables. Average pesticide consumption in vegetables in India around 0.678 a.i. kg/ha. Maximum pesticide usage in chilli followed by brinjal, cole crops and okra. Global agro-chemical consumption dominated by fruits and vegetables, accounting 25 % of total pesticide market. Major insect-pests of different vegetable crops and their management are as follows:

Cole crops

Cabbage (*Brassica oleracea var capitata*) and Cauliflower (*Brassica olercea var botrytis* L) are grown in cold and moist climate. It is rich in minerals namely iron, magnesium, phosphorous and sodium. Cole crops were attacked by number of insect pests and cause heavy economic loss. The major insect pests are described below along with their management:

Diamond back moth *Plutella xylostella* Linn. (Lepidoptera: Plutellidae)

Economic importance: This pest serious during February – March, and later in August –September months.

Marks of identification and life cycle: The moths are brown in color with conspicuous white spots on the fore wings, which appear like diamond patterns when the wings lie flat over the body. A female may lay 18-356 eggs in her life time yellowish eggs laid singly or in batches of 2-40 on the underside of leaves. Larvae are pale yellowish green with fine black hair scattered all over the body. The larval duration lasts for 9-17 days. Larva constructs a barrel shaped silken cocoon for pupation and is attached to the leaf surface. Pupal stage complete in 4-5days. The life cycle complete in 4-5 weeks.

Nature of damage: Caterpillars damage the leaves of cauliflower, cabbage and rape seed. In earlier stages, caterpillar feed in mines on the lower side of cabbage leaves and, in the later stages, feed exposed on the leaves. Sometime they produce shot holes in leaves. The growth of young plants is greatly inhibited. Central leaves of cabbage or cauliflower may be riddled and the vegetable rendered unfit for human consumption.

Diamond Back Moth, Adult and pupae

Leaf webber, *Crocidolomia binotalis* Z. (Pyraustidae: Lepidoptera)

Distribution: Present in all localities where cabbage and cauliflower are grown.

Nature of damage: The leaves are skeletonized by the larvae which remain on the under surface of leaves in webs and feed on them. They also attack flower buds and pods. Often it assumes serious proportions. It attacks cabbage, cauliflower, radish, mustard and other crucifers and the weed *Gynandropsis pentaphylla*.

Life history: The small moth with light brownish fore wings lays the eggs in masses, each mass containing 40 - 100 overlapping flat eggs. The incubation period ranges from 5 – 15 days depending on weather. The larva with red head has brown longitudinal stripes and rows of tubercles with short hairs on its pale violaceous body. It becomes full grown in 24–27 days during summer and in 51 days during winter. It pupates in an earthen cocoon and emerges as adult in 14-20 days.

Management strategies

(i) Mustard sown as trap crop twice i.e. 12 days preceding planting cabbage and again 40 days later controls DBM.

(ii) Spray of *Bacillus thurengensis* var Kurstaki @ 0.75 kg/ha or NPV @ 250 LE/ha

(iii) If still the infestation persists spray Emamectin benzoate 1.9% EC @ 1.5ml/l or Indoxacarb/Avant @ 0.5ml/l or Qinalphos @ 1.0ml/l or Spinosad 2.5% SC @ 1ml/l.

(iv) Waiting period must be followed after each spray. (Table 1)

Cabbage butterfly *Pieris brassicae* (Linnaeus) (Pieriidae: Lepidoptera)

Distribution: In India, it is widely distributed along the entire Himalayan region. *P. brassicae* is comparatively more common and destructive. It causes severe damage to cabbage, cauliflower, radish, turnip as also mustard and rape. The pest passes winter in the plains and migrates to hilly regions during summer. During September to April, it breeds on rape and mustard.

Nature of damage: On hatching, the young caterpillars feed gregariously on leaves for a couple of days and then disperse, spreading infestation to the adjacent plants and fields. As a result of their feeding the leaves are skeletonized, sometimes the caterpillars bore into the heads of cabbage and cauliflower.

Life history: Eggs are laid in clusters under surface of the leaf. A single female lays only 2-3 egg-masses of 50-80 eggs each. Eggs are flask-shaped, about one mm long and yellowish in colour. Full grown caterpillars are 38 - 44 mm long, velvety bluish-green in colour with black dots and yellow dorsal and lateral stripes covered with white hair. Pupae are yellowish-green with black spots and dots. Adult butterflies have snow-white forewings with black distal margins more developed in females than in males; hind wings are also pure white with black apical spots. Wing expanse is 60-70 mm. Moths emerging in summer are larger in size than those of winter. Incubation, larval and pupal periods are on an average 3.2, 5.6 and 7.3 days during May extending upto 17.6, 40.7 and 28.8 days respectively in January. Generally there are two generations during winter (plains) and 4-5 in summer (hilly region) (Atwal, 2005).

Butterfly

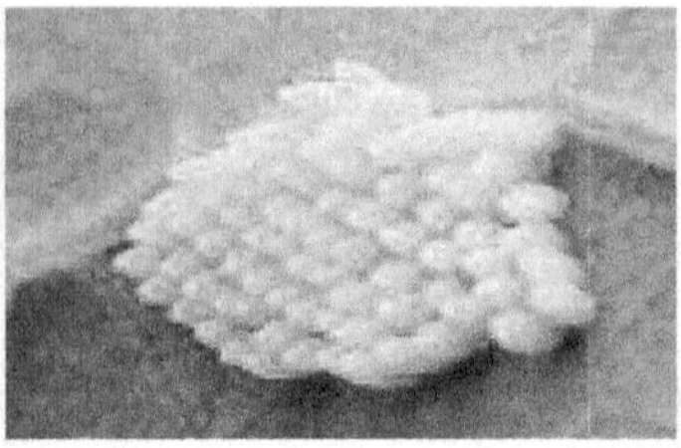

Eggs

larvae and damaged

Management strategies

(i) Pest can be checked by handpicking and mechanical destruction of caterpillars during early stage of attack when the caterpillars feed gregariously.

(ii) Use 5 per cent Neem Seed Kernel Extract (NSKE)

(iii) Use *Baccilus thurengensis* @ 0.75kg/ha or NPV @ 250LE/ha is effective.

(iv) If still the infestation persist spray Emamectin benzoate 1.9% EC @1.5ml/l or Indoxacarb 14.5% SC/Avant @ 1.0ml/l or Qinalphos @1.0 ml/l or Malathion @ 1.5ml/l or Spinosad 2.5% SC @1ml/l.

(v) Waiting period must be followed after each spray (Table 1).

Table 1: Waiting periods of different Insecticides (Reddy, 2010)

Name of Insecticide	Target Pests	Dose/ha	Waiting Periods (Days)
Thiamethoxam 25% WG	Aphids, Jassids, White fly, Thrips	250ml	7-10
Acetamiprid 20 %SC	Aphids, Jassids, Thrips	100 g	15
Emamectin Benzoate 5% SG	Borers, DBM	200 g	5-10
Novaluron10EC	Borers	750 g	5-10
Spinosad 2.5% SC	DBM /Borers	600g	5
Imidacloprid 70% WG	Aphids, Jassids	35g	7
Imidacloprid 17.8 SL	Aphids, Jassids	100 ml	40
Indoxacarb 14.5 %SC	DBM, Borers	300 ml	7
Imidacloprid 4 %FS	Seed treatment	500-900g/100kg	—
Fipronil 5 % SC	DBM, Borers	800-1000	7
Dicofol 8.5 % EC	Mites	2500 ml	15-20
Fenazaquin 10 EC	Mites	1000ml	7
Propargite 57 EC	Mite	1000 ml	7

Cabbage aphid *Brevicoryne brassicae* (Aphididae: Homoptera)

Distribution: Cabbage aphid *Brevicoryne brassicae* was originally confined to Palaearctic or Holarctic regions but at present it has a very wide range of distribution.

Nature of damage : Colonies of these insects are often found on tender shoots and as a result of sucking of vital sap from the tissues, the plant remain stunted in growth resulting in poor head formation. In the case of severe infestation plants may completely dry up and die away. When infestation occurs on seedlings, they lose their vigor, get distorted and become unfit for transplanting. The aphids also produce copious quantity of honeydew which makes the plants sticky and favors the growth of sooty mould, as a result a black coating is formed on affected plant parts hindering the photosynthesis and adversely affecting the plant growth. IPM provided better results as compared to other plots (Rahman.1992).

Life history: Reproduction is mostly viviparous parthenogenetic during summer and mild winter. However, during severe winter sexual reproduction may also occur. Eggs when present are pale yellow with greenish tinge. Nymphs are 1.0-1.5 mm long and yellowish-green in color while adults are 1.8-2.0 mm long and darker in color than nymphs. Eggs are laid during November – December. These hatch in 20-22 weeks.

The nymphs mature in about 2 weeks and immediately start producing young ones, without mating. A single female may produce 40-45 young ones during her life time. The life cycle is completed in 11-45 days and as many as 21 generations have been recorded during a year when provided with favourable conditions.

Management strategies

(i) As soon as aphid infestation appears, cut and destroy the infested shoots mechanically.

(ii) Yellow sticky trap is effective for managing aphids.

(iii) If still the aphid population persist spray Imidacloprid 70% WG @ 0.5ml/l or Acetamiprid 20% SP @ 0.2g/l or Thiamethoxam 25 % WG @ 0.2g/l or Metasystox @ 1.5ml/l.

(iv) Waiting periods must be followed for every insecticides.

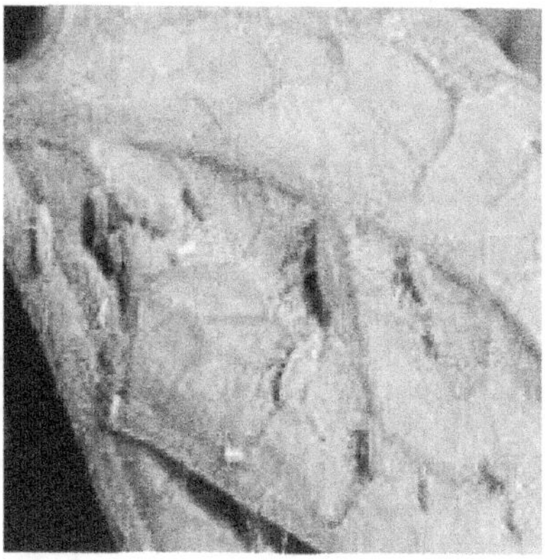

Aphid, *Brevicoryne brassicae*

Mustard Sawfly, *Athalia lugens proxima* Kulg (Tenthredinidae: Hymenoptera)

Distribution: Mustard sawfly is one of the very few hymenopterous insects reported as crop pests, and that too with chewing and biting habits. It is a cold weather pest found all over the Indian sub-continent. It is a major pest of not only cole crops but of almost all cruciferous plants, including rape and mustard. The peak period of activity is during September to December after which the activity declines; the pest is hardly noticed from March to July and appears on radish by the end of July.

Nature of damage: Eggs are laid singly, mostly during day time and inserted into leaf tissues near the periphery of leaves. On hatching the grubs nibble the margins of tender leaves but later on bite holes in the leaves. Grubs are diurnal in habit and feed generally during early morning and evening hours. With slight disturbance they fall on the soil and feign death. (Atwal, 2005).

Life history: A female lays on an average 35 eggs (20-150). Egg period is 6-8 days. Newly hatched grubs are 2-3 mm long, smooth, cylindrical and greenish-grey in colour; full grown ones are cylindrical in shape, 16-20 mm long and greenish-black in colour. They look and behave like caterpillars but have 8 pairs of forelegs. Grub development takes 21-31 days. Adults are 8-12 mm long, having dark head and thorax, orange coloured abdomen and translucent smoky wings with black veins. Females have a strong saw-like ovipositor – hence it has been given the popular name sawfly. They generally do not fly long distances but hop from leaf to leaf or fly from one plant to another plant. Their activity is pronounced during days while the insects remain practically motionless at night. Pre-pupal and pupal periods last for 3-4 and 7-10 days respectively. Severe winter is passed in pupal stage and lasts for about 14 weeks. In Northern India there are three generations during cold season. In South India where there is no severe winter, the pest undergoes as many as 10 overlapping generations in a year.

Management strategies

(i) Hand-picking of grubs which are not active during dawn and dusk if the area under crop is limited.

(ii) Use Imidacloprid 17.8 SL @ 0.5ml/l or Acetamiprid 20 % SP @ 100g/ha or Thiamethoxam 25 % WG @ 100g/ha at seedling stage.

(iii) Waiting periods must be followed for every insecticides (Table1).

Painted Bug, *Bagrada cruciferarum* Kirkaldy (Pentatomidae: Hemiptera)

Distribution: Painted bug *Bagrada cruciferarum* and *B. hilaris* (Burmester) have been recorded as major pests of various *Brassica* spp. as also other cruciferous crops and weeds. *B. cruciferarum* has been reported from East Africa, Afghanistan, Pakistan, Sri Lanka, India and South-east Asia while *B. hilaris* is found in East, West and South Africa, Italy, Iran, Iraq, Pakistan, India, Sri Lanka and USSR. The adults appear in field around October and their activity decreases with the onset of summer but is again accelerated in autumn.

Nature of damage : Both nymphs and adults suck cell sap from tender plant parts causing yellowing of leaves which gradually dry up and ultimately fall down exposing the plants to secondary invasion of bacteria and fungi. The plants wilt and wither affecting adversely the yield both quantitatively and qualitatively.

Life history: Eggs are laid singly or in batches of 2-12 on leaves, stems and flower buds. These are oval in shape about one mm long, pale yellow when freshly laid gradually becoming pinkish-orange. Nymphs are beautifully patterned with a mixture of black, white and orange colour, 1.5-4.5 mm long depending on their age. Adults are also black and orange colour bugs similar in colour pattern as nymphs – that's why they have earned the common name of painted bugs. Males are 6-7 mm long and females 7-8 mm. The mating takes place 2-6 days after the final nymphal moult and the oviposition commences a week after first mating and may continue intermittently throughout the life span of the female. A single female may lay as many as 230 eggs @ 15-20 eggs per day. Eggs and nymphal duration is recorded as 2-5 and 18-20 days respectively. A single life-cycle is completed in 3-4 weeks and adults live for 16-18 days with 6-8 generations in a year.

Management strategies

(i) Clean cultivation by removing weeds harbouring this pest is imperative for avoiding infestation of these bugs.

(ii) If still the aphid population persists spray Imidacloprid 70 % WG @ 0.5ml/l or Acetamiprid 20%SP @ 0.2g/l or Thiamethoxam 25 % WG @ 0.2g/l or Metasystox @ 1.5ml/l.

(iii) Waiting periods must be followed for every insecticides (Table 1).

Tobacco caterpillar *Spodoptera litura* Fabricius (Lepidoptera: Noctuidae)

Economic importance: Polyphagous in nature and serious during August-October and February–March.

Marks of identification and life cycle: The female adult lays about 300 eggs in clusters. These clusters are covered over by brown hair and they hatch in about 3-5 days. Larva is velvety black with yellowish-green dorsal stripes and lateral white bands. They pass through six stage lasts 7-15 days. The life cycle is complete in 32-60 days and the pest completes eight generation in a year.

Nature of damage: The larvae feed gregariously for the first few days and then disperse to feed individually. The larvae feed on leaves and fresh growth of the plant.

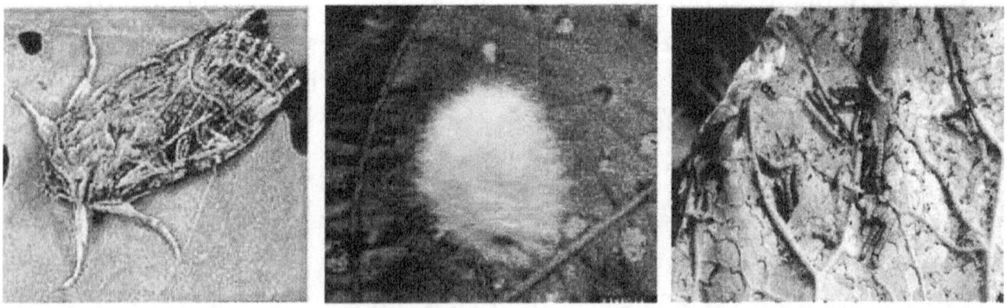

Adult Eggs Larvae and damaged

Cabbage Borer, *Hellula undalis* Fabr. (Lepidoptera: Pyralidae)

Economic importance: This pest is distributed in all over country.

Marks of identification and life cycle: The female moths lay eggs singly but more often in clusters, on the under surface of the leaves or some other parts of the plant. The caterpillar feed in the heart of the cabbage and become full-grown in 7-12 days. Pupal period covers in 6-10 days.

Nature of damage: The larvae bore into the central shoots and the plant is unable to bear the flower head. Since the attack is mostly on young plants in the nursery and the fields. The caterpillar first mine into the leaves. Later on, they feed on the leaf surface, sheltered within the silken passages. As they grow bigger, they bore into the heads of cauliflower and cabbage. They prevent head initiation causing multiple shoots or heads.

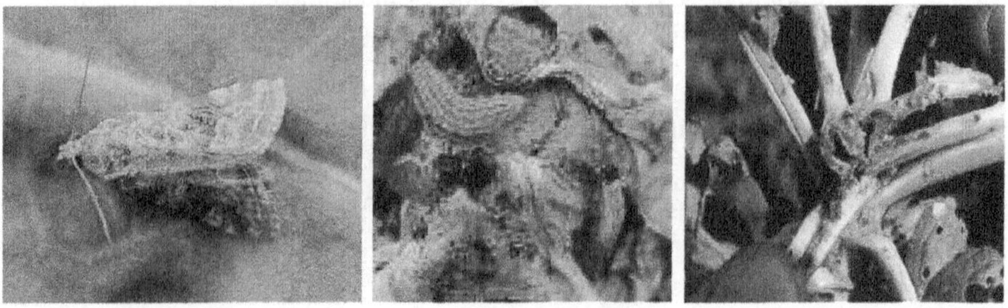

Adult Larvae Damaged plant

Management strategies to be followed for all the cole crops pests

- Growing of recommended varieties.
- Eliminate weeds and previous crop residue.
- Timely sowing of nursery and transplanting of seedling.
- Screening of seedling in nursery beds.
- Intercropping of one row of tomato (30 days after cabbage) and one row of cabbage reduced the DBM and leaf webber incidence.
- The practice of trap/intercropping of mustard in cruciferous vegetables may be helpful in lowering the pest status of DBM.
- Use light traps for adult DBM @ 3traps/acre. Hang a bulb over a bucket of water. Within 3-4 days most of the adults get trapped and killed.
- Removal and destruction of the egg masses and clusters of larvae of cabbage caterpillar and tobacco caterpillar.
- Introduction of *Apanteles plutellae* (Kurdyumov) 1000 adults release-1 every two-week interval upto harvest helps to check DBM population. For cabbage butterfly release *Apanteles glomeratus*. Release of pentatomid predator, *Eocanthecona furcellata* (Wolf) were found to feed voraciously on *Crocidolomia* and *Hellula* spp.
- Releases of *Chrysoperla* @5larvae/plant+5% neemazal+mechanical collection + planting of mustard crop on the boarder+ release of *T. pieris* @1,00,000/ha+spray of Delfin WG @300gm/acre (for cabbage caterpillar, Kaur and Virk, 2015)
- Microbials derived from actinomycetes, bacteria and fungi, viruses are found to be very effective against these pests
- Spray *B.t* (1g/litre) if DBM 1.0/plant is observed.
- Spray NSKE 5 % at 10-15 days interval. Maximum 3-4 sprays are required
- The use of overhead sprinkler system helps establishment of larval parasitoid, *A. plutellae* for control of DBM.

Chemical control

- Spot application of insecticide in case of tobacco and cabbage caterpillar where less attack in the field.
- Repetition of the spray after 10 days, if necessary.

Insect	Insecticides	Dose
Head borer	Hexavin 50 WP	150g
Diamond back moth,	Dipel	300ml
Tobacco caterpillar and	Halt	300g
Cabbage caterpillar	Emamectin benzoate SG	70g
	Indoxacarb 15.8 EC	130ml
	Cartap hydrochloride 50 SP	200g
	Spinosad 2.5 SC	250ml
	Quinalphos 25 EC	200-400ml
Tobacco caterpillar	Novaluron 10 EC	150ml

References

Annonymous 2015. Package of practice for cultivation of vegetables, Punjab Agricultural University, Ludhiana.

Anonymous, 2015. www.plantnatural.com/tomato-gardening guru. 6.7.15.

Devi, L.L., Senapati,A.K. and Chatterjee, M.L. 2015. Biorational management of tomato fruit borer, Helicoverpa armigera at new alluvial zone of west Bengal. In 4th congress on insect science "Entomology for sustainable Agriculture .April 16-17 at PAU, Ludhiana

Dhandapani, N., Umeshchandra, Shelkar, R. and Murugan, M. 2003. Bio-intensive pest management (BIPM) in major vegetable crops: An Indian perspective. Food, Agriculture & Environment Vol.1 (2): 333-339.

Gayatridevi, S. and Giraddi, R.S. 2009. Effect of date of planting on the activity of thrips, mites and development of leaf curl in chilli (*Capsicum annum* L.) *Karnataka Journal of Agricultural Sciences* 22 (1):206-207.

Ghosh,A., Chatterjee, M. and Roy, A. 2010. Bio-efficacy of spinosad against tomato fruit borer (*Helicoverpa armigera* Hub.) (Lepidoptera: Noctuidae) and its natural enemies. *Journal of Horticulture and Forestry* 2(5):108-111.

Gundannavar, K.P., Giraddi, Kulkarni, K.A. and Awaknavar, J.S. 2007. Development of integrated pest management modules for chilli pests. *Karnataka Journal of Agricultural Sciences* 20 (4):757-760.

ICRISAT. 2011. Integrated Pest Management in Chilli, Tomato, Onion and Potato crops.PP-4.

Kaur, R. and Virk, J.S. 2015.development of biocontrol based IPM module against cabbage butterfly, *Pieris brassicae* (Linnaeus) on cauliflower. In 4th congress on insect science "Entomology for sustainable Agriculture .April 16-17 at PAU, Ludhiana.

Krishnamoorti, A. and Mani, M. 1988. Feasibility of managing *H. armigera* on tomato through parasitoid. In: National Workshop on *Heliothis* management. Feb. 18-19. TNAU, Coimbatore.

Rahman, M.M., Rahman, S.M.M. and Akter, A. 2011. Comparative performance of some insecticides and botanicals against chilli fruit borer. *Journal of Experimental Sciences* 2(1): 27-31.

Rai, A.B., Loganathan, M. Halder, J., Venkataravanappa, V. and Naik, P.S. 2014. Eco-friendly Approaches for Sustainable Management of Vegetable Pests. IIVR Technical Bulletin No. 53, IIVR, Varanasi, pp. 104.

Rajavel, D. S., Mohanraj, A. and Bharathi, K. 2011.Efficacy of chlorantraniliprole (Coragen 20SC) against brinjal shoot and fruit borer, *Leucinodes orbonalis* (Guen.). *Pest Management in Horticultural Ecosystems*, 17(1): 28-31.

Reddy, M.R.S. and Reddy, G.S. 1999. An eco-friendly method to combat *Helicoverpa armigera* (Hub.). *Insect Environ.* 4: 143-144.

Sardana, H.R. and Sabir,N. 2007. IPM strategies for tomato and cabbage. Extension folder by NCIPM, New Delhi.

Sardana, H.R., Tyagi, A. and Singh, A. 2005. Knowledge resources on fruit flies (Tephritidae: Diptera) in India. National center for integrated pest management, New Delhi, India.

Tatagar M. H., Mohankumar H. D., Shivaprasad M. and Mesta R. K. 2009. Bio-efficacy of flubendiamide 20 WG against chilli fruit borers, *Helicoverpa armigera* (Hub.) and *Spodoptera litura* (Fb.) *Karnataka J. Agric. Sci.* 22(3-Spl. Issue): 579-581.

Tatagar, M.H, Awaknavar, J.S., Giraddi, R.S., Mohankumar, H.D., Mallapur, C.P. and Kataraki, P.A. 2011. Role of border crop for the management of chilli leaf curl caused due to thrips, *Scirtothrips dorsalis* (Hood) and mites, *Polyphagotarsonemus latus* (Banks). *Karnataka Journal of Agricultural Sciences* 24 (3):294-299.

6

Insect Pests Infesting Major Vegetable Crops and Their Management Strategies - II

R. M. Srivastava, S.K. Maurya, Tanuja Phartiyal and Amandeep Kaur

Okra *Abelmoschus esculentus* L.

Cotton Jassid *Amrasca biguttula* Ishida (Hemiptera: Cicadellidae)

Economic importance: Sucking pest cause damage during May to September. Adults are also found on plants such as potato, brinjal, tomato, etc.

Mark of identification and life cycle: Adults are about 3 mm long and greenish yellow during the summer, acquiring a reddish tinge in the winter. The winged adults jump or fly away at the slightest disturbance and are also attracted to light at night. Females lay about 15-38 yellowish eggs on the underside of the leaves, embedding them into the leaf veins. The eggs hatch in 4-5 days. Nymphs are wedge-shaped and are very active and complete five instars with 3-5 days in each instars. Adult live for 11days.

Nature of damage: Damage to the crop is caused by the adults as well as by the nymphs, both of which are very agile and move briskly, forward and sideways. They suck cell-sap from the underside of the leaves and pass through six stages of growth in 7-21 days. They feed constantly on the plant juice. The pest completes seven generations in a year. Damage symptoms: Injury to plants is due to the loss of sap and probably also due to the injection of toxins. The attacked leaves turn pale and then rust-red. With change in appearance, the leaves also turn downwards, dry up and fall to the ground.

Management strategies

- Tolerant/less susceptible varieties. Tolerant varieties to some important pests are, Punjab-8, Punjab-7, Punjab Padmini against *Amrasca kerri*; Parbhani Kranti, Varsha Uphar, Arka Anamika, IC-7194, and IC-13999 for Yellow Vein Mosaic.
- Seed treatment with Imidacloprid (Gaucho) 70 WS @ 3 g/kg and Thiamethoxam (Dhawan *et. al*., 2011)
- Conservation of parasitoids is very important as parasitoids like *Erythmelus empoascae, Stethymium empoascae*, *Chrysoperla* and spiders are found to be active against *Amrasca bigutula bigutula*.
- Spraying once or twice at fortnightly interval with 560ml malathion 50 EC or confidor 200SL @ 250ml in 100-125liters of water per acre.
- As soon as flowering starts give three sprays at 15 days interval of 70 gm of proclaim 05 SC (emamectin benzoate), 100ml of sumicidine 20 EC (fenvalerate) or 80 ml of cymbush 25 EC (cypermethrin) in 100-125 liters of water/acre.

Okra Shoot and Fruit borer, also called spotted bollworm *Earias vittella* (Fabricius) (Arctidae: Lepidoptera)

Economic importance: Polyphagous in nature and cause damage during May to September.

Distribution: It is widely distributed, recorded from Pakistan, India, Sri Lanka, Bangladesh, Burma, Indonesia, New Guinea, and Fiji. It is an oligophagous pest having okra and cotton as its main hosts. It is also found feeding on a large number of malvaceous plants, both wild as well as cultivated.

Mark of identification and life cycle: The moth is 20-25 mm across the wings. The adult moth of *E. insulana* possesses green/yellowish green forewings; whereas hind wings appear pale. The moth of *E. vittella* has green forewings marked with green streak running medially from base to the outer margins. Egg: Spherical in shape and bluish in colour. Eggs are spherical in shape, about half mm in diameter, light bluish-green in colour and beautifully sculptured with 26 - 32 longitudinal ridges; the alternate ridges project upwards to form a crown. The moths emerge at dusk; mating takes place 2-3 days after emergence and oviposition commences after 1-5 days of mating. A female moth lays eggs (150-250 per female) singly on flower buds, branches and young leaves. Hairy part of plant is preferred for egg laying. The larva is stout; spindle shaped bears two pairs of fleshy, finger like tubercles. Though the colour is slightly variable, yet *E. insulana* has cream colour body marked with orange dots on prothorax. *E. Vittella* is brownish in colour possessing medium longitudinal stripe but without finger shaped processes. Pupa is enclosed in tough silken cocoon, which are dirty white or light brown colour. Pupation occurs either on plants or in the fallen leaves. Incubation, larval and pupal periods last for 3-9, 9-20 (50-60 during winter) and 8-12 days respectively. A single life-cycle takes 22-25 days extending up to 74 days during winter and there may be 8-12 generations in a year. Another fruit borer, which is also known as spotted bollworm of cotton *Earias insulana* (Boisduval), is found damaging okra especially in drier regions.

Nature of damage: Eggs are usually laid singly on buds and flowers and occasionally on fruits as well, but in absence of these parts i.e. at early stage of crop growth, the eggs are laid on shoot tips. When the crop is only a few weeks old, the freshly hatched larvae bore into tender shoots and tunnel downwards, these shoots wither, droop down and ultimately the growing points are killed, side shoots may arise giving the plants a bushy appearance. With the formation of buds, flowers and fruits, the caterpillars bore inside these and feed on inner tissues. They move from bud to bud and fruit to fruit thus causing damage to a number of fruiting bodies. The damaged buds and flowers wither and fall down without bearing any fruit whereas the affected fruits become deformed in shape and remain stunted in growth.

Adult

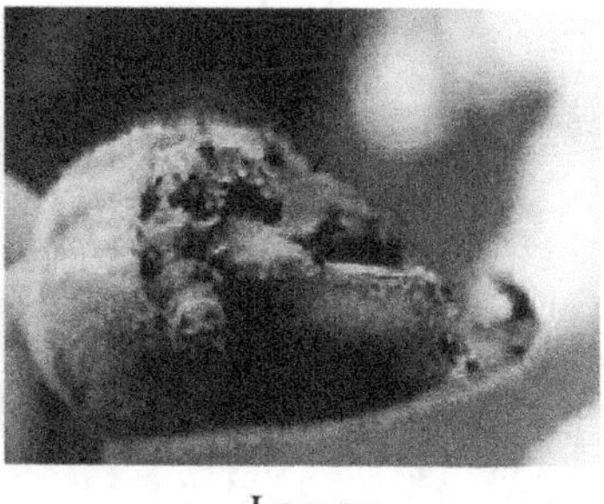

Larvae

Damaged fruit

Management strategies

- Tolerant/less susceptible varieties. They reduce the pest incidence by non-preferance, antibiosis and/or tolerance. Tolerent varieties to some important pests are, Punjab-8, Clemson Spineless, MP-7, AE-57 for *E. vitella.*
- Deep ploughing to expose resting pupae.
- Uproot hollyhock and rationed cotton, which are host plants.
- Avoid monocropping of bhindi year after year.
- Remove regularly the attacked fruits and burry deep in the soil.
- Remove debris and all the alternate host plants from field; collect and destroy all the infested shoots and fruits.
- Release egg parasitoid *Trichogramma chilonis* Ishii, *T. brasilensis*, and larval parasitoids *Chelonus blackburni* or *Bracon brevicornis* or *Apanteles* sp. at 35 to 70 days.
- Commercial formulations of *Bacillus thuringiensis var. kurstaki Viz.* Biolep followed by Bioasp proved to be most effective at all concentrations against *E. vitella*
- Spraying once or twice at fortnightly interval with 560ml malathion 50 EC in 100-125 liters of water per acre.
- Spray Neem Seed Kernal Extract @ 5 ml/l after flowering.
- Use Fipronil 5 %SC @ 800-1000 ml/ha or Novaluron 10EC @750ml/ha or Spinosad 45 %SC @ 160g/ha or Emamectin benzoate 5%SG @ 200g/ha.
- As soon as flowering starts give three sprays at 15 days interval of 70 gm of proclaim 05 SC (emamectin benzoate), 100ml of sumicidine 20 EC (fenvalerate) or 80 ml of cymbush 25 EC (cypermethrin)in 100-125 liters of water/acre.

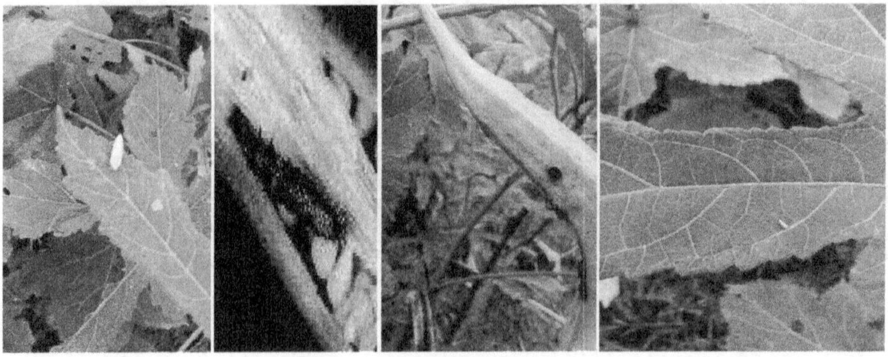

1. Leafhopper 2. Fruit borer 3. Damage of borer 4. White fly

2. Whitefly, *Bemisia tabaci* (Aleyrodidae: Hemiptera)

Distribution: Commonly known as cotton whitefly is found in most of the countries in tropics and subtropics. Its main hosts are cotton, tobacco and some winter vegetables; including tomato, the infestation on these crops is sporadically severe.

Nature of damage : White, tiny, scale-like insects may be seen darting about near the plants or crowding in between the veins on ventral surface of leaves, sucking the sap from the infested parts. The pest is more active during the dry season and its activity decreases with the onset of rains. As a result of their feeding the affected parts become yellowish, the leaves wrinkle and curl downwards and are ultimately shed. Besides the feeding damage, these insects also exude honeydew which favours the development of sooty mould. In case of severe infestation, this black coating is so heavy that it interferes with the photosynthetic activity of the plant resulting in stunted growth. This whitefly also acts as a vector, transmitting the Yellow Vein Mosaic Virus (YVMV) in okra.

Life history: Eggs are pear-shaped, light yellowish in colour, about 2 mm long and can be seen standing upright on leaves, being anchored by a tail-like appendage inserted into the stoma of leaves. On hatching, the nymphs crawl a little, settle down on a succulent spot on the same leaf and never move again during that stage. Nymphs are oval, scale-like and greenish-white in colour. Adults are minute insects, about one mm long, covered completely with a white waxy bloom. Incubation period is 3 - 5 days in summer extending up to 33 days during winter. Nymphal development takes 9 - 14 and 17 - 81 days in summer and winter, respectively and pupal period lasts for 2 - 8 days being longer during winter than in summer. A life-cycle may be completed in as little as 14 days or it may even be prolonged up to 107 days. There are about 12 overlapping generations in a year.

Management strategies

(i) Crop rotation is effective tool to prevent pest population. Not use same group of crop in same field for a long time.

(ii) Sticky trap is effective to control pests population.

(iii) Use Imidacloprid 17.8 SL @ 0.5ml/l or Acetamiprid 20%SP @ 100g/ha or Thiamethoxam 25 %WG @ 100g/ha or Metasystox @ 1.5ml/l within one month of transplanting. Single white fly may transmit the viral disease.

3. Red spider mite, *Tetranychus cinnabarinus* (Boisduval) (Tetranychidae: Arachnida)

Distribution: *Tetranychus cinnabarinus*, commonly called red spider mite, is the most common and destructive species attacking okra all over India. It has a worldwide distribution and is highly polyphagous having a very wide range of host plants.

Nature of damage: Colonies of mites comprising of eggs, nymphs and adults are found feeding on ventral surface of leaves under protective cover of fine silken webs. As a result of their feeding innumerable yellow spots appear on the dorsal surface of leaves and the affected leaves gradually start curling and finally get wrinkled and crumpled. This in turn affects the growth and fruit formation capacity of the plants.

Life history: Eggs are globular in shape, about 0.1 mm in diameter and whitish in colour. Larvae are about 0.2 mm in length and pinkish in colour. Nymphs are greenish-red in colour and about 3 mm in length. Larvae and nymphs look alike in shape but can be easily distinguished as larvae have 3 pairs of legs while nymphs and adults have 4 pairs of legs. There are only two nymphal stages – protonymphal and deutonymphal. Adults are ovate in shape, reddish-brown in colour and 0.4 mm (male) - 0.5 mm (female) in length with four pairs of legs. Eggs hatch in 4 to 7 days; larval development takes 3 - 5 days; protonymphal and deutonymphal stages last for 3 - 4 days each. Longevity of adult males and females is 4 - 9 and 9 - 18 days respectively. The females that are active during summer in northern India become active with the onset of monsoon and lay eggs parthenogenetically. These unfertilized eggs give rise to males only but the subsequent generations are sexual.

Management strategies

(i) Propargite 57% EC @ 1200 ml/ha or Fenpyroximate 5% EC @ 500ml/ha.

Cucurbits

Cucurbit Fruit Fly *Bactrocera cucurbitae* (Diptera:Tephritidae)

Economic importance: Fruit flies are most damaging for cucurbitaceous crops. Preferred hosts are cucurbits (squash, melon, etc.). The infestation varies from 50-100 per cent in different cucurbitaceous crops. Other hosts include solanaceous plants (tomato, eggplant, pepper, etc.) and papaya. 30-40% damage due to fruit flies in vegetable and fruit crops. Economic losses worth Rs. 7000 crore has been reported to occur annually as a result of fruit flies infestation in India (Sardana, 2005).

Mark of identification and life cycle: The female fly makes a cavity with the help of its sharp ovipositor and inserts the white, cigar-shaped, slightly curved eggs singly or in groups into the flowers and tender fruits. The incubation period is 3-9 days. The freshly hatched maggots bore into the fruit pulp by forming serpentine galleries and contaminating them with its frass and providing entry points for saprophytic fungi and bacteria, which cause the rotting of the fruit. Due to feeding on pulp, there is premature dropping of fruits and make them unfit for consumption. Adult flies are medium sized (4-5 mm long) with reddish brown and their wings are hyaline with a dark patch on the outer margins. Hind cross veins thickened with brown and grey spots at the apex. The full-grown maggots come out of the fruits and drop to the ground and pupate in the soil. The total life cycle occupies 14-34 days depending on weather conditions. There are several over lapping generations in a year.

Nature of damage: Adult female flies select soft and young fruits for oviposition by puncturing the rind with the ovipositor. Such damaged fruits show signs of raised and brown resinous encrustation at the sites of ovipositional punctures due to discharge and drain out fruit juice through punctures. Infested fruits are either rotten or deformed in shape due to the microbial infection. Such rotten or deformed fruits are not fit for sale or human consumption. Infested fruits with apparent firmness and growth show internal decay and foul smell when cut open.

Fruit fly

Management strategies

- Growing of resistant or early maturing varieties.
- Collect the damaged fruits and destroyed them by burying deep in soil.
- Use of sex pheromone like "Cuelure".
- Fruit flies can be controlled by using poison bait (100g smashed ripened pumpkin + 0.5g Dipterex + 100ml water).
- Apply the bait spray contacting 0.05% malathion, 1% gur/sugar (200g gur/sugar+20 ml malathion+20 liters of water). Repeat the spray at weekly intervals when the attack is high.
- Spraying the bait on the lower surface of the leaves of maize plants grown in rows at distance of 8-10m as trap crop has been found to be effective. (Anonymous 2015).

Red Pumpkin Beetle *Raphidopalpa foveicollis* Lucas (Coleoptera: Chrysomelidae)

Economic importance: The red pumpkin beetle is a serious pest of cucurbitaceous plants, except bitter gourd.

Mark of identification and life cycle: adult weevils wake up after hibernation in the early March. After mating a female lay egg's singly or in batches of 8-9 in the moist soil at the base of the host plant. As many as 300 eggs are laid by a single female. Eggs are elongated and brown in colour. Egg hatches into larva in 6-15 days (5-8 days in optimal conditions). The whitish grub with brown head bores and feed upon plant roots, fallen leaves and fruits lying on the surface of soil. Grub moults four times during 13-25 days of their larval period. Moulting occurs inside the soil. A fully grown grub moves deep into the soil (1.3-25.4 cms deep) and pupate within a water-proof, thick walled oval cocoon. Pupation period lasts for 7-17 days, after which it metamorphoses into adult beetle. The adult comes out of the soil to feed upon the host plants and to breed. After about seven days of emergence beetle start laying eggs. Five generations are completed during March to October. A complete life cycle takes about 25-37 days. The adults hibernate in November inside soil or among dry weeds and appear again in March.

Nature of damage: Adult beetles feed voraciously on the leaf lamina by making irregular holes. The maximum damage is done when the crop is in the cotyledon stage. The first generation is therefore more injurious than the subsequent generations. The adult insect also feed on the leaves of grown up plant by scrapping off their chlorophyll and make the leaves net like appearance. The attacked plants may wither away and re-sowing of the crop may become necessary in certain cases. The larvae cause damage in various ways by boring into the roots and the underground stem portion and by feeding on the leaves and fruits line in contact with the soil. The damaged roots and the underground roots and the underground stems portion may rot due to infection by the saprophytic fungi. The young and smaller fruits of the infested creepers may dry up, whereas the bigger and mature fruits become unfit for human consumption.

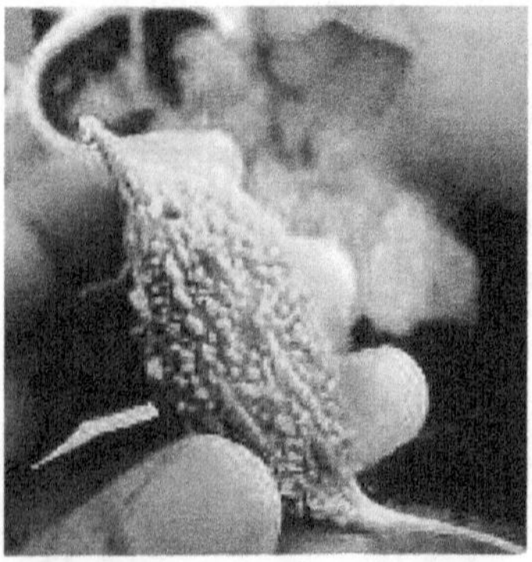

Management strategies

- The pest population can be suppressed effectively and its infestation can be squeeze in the bud, by regular killing and picking of eggs, grubs and adults, if the cropped area is small.
- Also the larvae and adults can be shaken down in container of kerosinized water early in the morning.
- Just after germination 2.75 kg of furadon 3 G (carbofuran) /acre 3-4cm deep in soil near the bases of the plants and apply irrigation.
- The pest can be killed by spraying Malathion/ Fyfanon/ Zythiol 50 EC @2 ml of water. The treatment should be applied as soon as the pest appears and repeated at 15 days interval.
- Few scattered plants should be grown early in the season. They should be treated with strong insecticidal spray. So, that the adults attracted towards plant will die and the subsequent will have pest free crop.
- Collection and destruction of pest. In the early hours of the morning the beetles remain sluggish. They can be collected by hand nets and killed in kerosene oil.
- The pest gets repelled by ash or mixture of ash and insecticidal dust.
- The soil around the root of the plant should be sprayed with strong insecticides so that the developing grubs and pupa die before hatching. The pesticides used are malathion (0.05%) or the plant can be dusted with pyrethrum (5%).

Snake gourd semilooper, *Anadevidia (=Plusia) peponis* F. (Noctuidae: Lepidoptera)

Distribution: Found to occur in all localities where snake gourd is grown.

Nature of damage: The caterpillar cut the leaf partly and rolls it and lives inside the roll. It is a specific pest of snake gourd and the larvae defoliate the plants considerably if infestation is serious.

Life history: The brownish moth has shiny brown fore wings. The female moth lays white spherical eggs on the under surface of leaves. The semilooper caterpillar with humped last abdominal segment, measures 35 – 40 mm long. Its body is greenish with white longitudinal lines and black tubercles with thin hairs arising on them. It pupates in a thin silken cocoon in leaf fold. The pupa is greenish but turns dark brown before the emergence of the adult moth. Egg, larval and pupal periods last for 4-5, 24-30 and 7-8 days respectively. Its life-history occupies about six weeks.

Management strategies

(i) The larvae when found in small numbers may be hand-picked and destroyed.

***Epilachna* beetle: same as brinjal** – Please refer Chapter V.

Onion

Onion thrips, *Thrips tabaci* Hood (Thripidae: Thysanoptera)

Distribution: It is a polyphagous pest having a wide range of host plants distributed throughout India. It is also known to infest tea, grapevine, castor, cotton, *Prosopis juliflora* etc.

Nature of damage: Eggs are laid on or just under leaf tissues. Both nymphs and adults lacerate the leaf tissues and imbibe the oozing sap; sometimes even the buds and flowers are attacked. Tender leaves and growing shoots are preferred. The infested leaves start curling and crumbling and is ultimately shed whereas buds become brittle and drop down. Yield loss due to thrips attack may range from 25-50 %.

Life history: Eggs are minute and dirty white in colour. Nymphs and adults are also tiny, slender, fragile, and yellowish-straw in colour; adults have heavily fringed wings that are uniformly grey in colour. Reproduction is both sexual and parthenogenetic. In case of sexual reproduction, oviposition period lasts for about a month during which a female lays on an average 100 eggs @ 2 - 4 eggs per day. A single life-cycle is completed in 2 -2½ weeks as many as 25 overlapping generations in a year.

Management strategies

(i) The thrips *Franklinothrips vespiformis* (Crawford) and *Erythrothrips asiaticus* R. &. M. are predaceous on this thrips in nature and their population may be encouraged by avoiding chemical sprays.

(ii) Sticky trap is effective for controlling this trap.

(iii) Imidacloprid 17.8 SL @ 0.5 ml/l or Acetamiprid 20 %SP @ 100g/ha or Thiamethoxam 25 %WG @ 100g/ha. 15-30 days after transplanting. Fipronil 5%SC @ 2g/l or spinosad 45%SC @ 160g/ha may be use if thrips population exceed over 5 thrips/ plant.

Onion thrips

Radish

Aphid, *Aphis gossypii* Glover and *Myzus persicae* (Sulzer) (Aphididae: Homoptera)

Distribution: Both the species occur in all places in all seasons. The incidence is more in cool and humid seasons. *A. gossypii* is also found attacking cotton, bhindi, chillies and guava. This pest infest reddish crop at flowering time and reduce seed production.

Nature of damage: Both nymphs and adults are found in large number sucking the cell sap from leaves and tender apical shoots. The under surface of the leaves get crinkled and slightly curled backwards. The vitality of the plant is diminished and the plants turn yellow, get deformed and dry away. Besides this direct damage they also secrete copious quantity of honeydew on which sooty mould grows rapidly covering the affected parts with a thick black coating, which interferes with the photosynthetic activity of the plants. The infested plants become weak, pale and stunted in growth which consequently results in reduced fruit size.

Life history: Nymphs of *A. gossypii* are greenish-brown or yellowish in colour while adults are yellowish-green to dark green in colour, little over one mm in length and have a pair of siphunculi in the posterior side of abdomen. Wings when present are transparent with black veins. *A. gossypii* breeds during winter on a number of vegetables including brinjal from where it migrates in the month of April to melons and by June end return to cotton. Reproduction in case of *A. gossypii* is parthenogenetic viviparous and rate of multiplication is often phenomenal.

Adults of *M. persicae* are usually of green colour but may be pale brown to pinkish, 1.5 - 2.5 mm long with long clavate siphunculi. *M. persicae* reproduces by parthenogenetical viviparity during summer, monsoon and autumn seasons but sexually in cooler regions during winter. Adults normally perish due to severe cold and eggs overwinter in cracks and crevices on the bark of various temperate fruit trees. When the temperature rises, the eggs hatch and nymphs start feeding on blossoms. They mature in 3-4 days and reproduce parthenogenetically producing young ones which develop into wingless adults. After 2 - 3 generations when the temperature rises further or when there is too much crowding of these aphids, the winged forms are produced and these migrate to other crops including brinjal; again with fall in temperature, they migrate back to temperate fruit trees.

Management strategies

(i) Conservation of the coccinellids and syrphids that are found to feed on the aphids will reduce the numbers considerably without any insecticidal spray.

(ii) Yellow sticky trap is effective for controlling aphid population.

(iii) Imidacloprid 17.8 SL @ 0.5 ml/l or Acetamiprid 20%SP @ 100g/ha or Thiamethoxam 25%WG @ 100g/ha.

References

A.K Dhawan, Sandeep Kaur, Amandeep Kaur and Gurpreet Kaur. 2011. Management of cotton jassid, *Amrasca biguttula* (Ishida) through seed treatments in okra. *Indian Journal of Agricultural Sciences* 81(2): 192-193.

Annonymous 2015. Package of practice for cultivation of vegetables, Punjab Agricultural University, Ludhiana.

Anonymous, 2015. www.plantnatural.com/tomato-gardening guru. 6.7.15

Dhandapani, N., Umeshchandra, Shelkar, R. and Murugan, M. 2003. Bio-intensive pest management (BIPM) in major vegetable crops: An Indian perspective. Food, Agriculture & Environment Vol.1 (2): 333-339.

ICRISAT. 2011. Integrated Pest Management in Chilli, Tomato, Onion and Potato crops.PP-4.

Pandey, N.K., Ojha, P.K., and Upadhyay, K.D. 1999. Studies on the bioefficacy of *B.thuringiensis* formulations against *E. vitella* Fabr. Under laboratory conditions. RAU. J. Res. **9**(1): 57-61.

Rai, A.B., Loganathan, M. Halder, J., Venkataravanappa, V. and Naik, P.S. 2014. Eco-friendly Approaches for Sustainable Management of Vegetable Pests. IIVR Technical Bulletin No. 53, IIVR, Varanasi, pp. 104.

Reena and Ram Singh. 2007. Insecticidal potential of Pongamia pinnata seed fractions of methanol extract against Earias vittella (Lepidoptera: Noctuidae). Entomologia Generalis. 30(1): 051-062.

Sardana, H.R., Tyagi, A. and Singh, A. 2005. Knowledge resources on fruit flies (Tephritidae: Diptera) in India. National center for integrated pest management, New Delhi, India.

7

Insect Pests Infesting Major Vegetable Crops and Their Management Strategies - III

Amandeep Kaur, R.M. Srivastava, S.K. Maurya, Tanuja Phartiyal and Reena

Solanaceous Vegetables

Solanaceous vegetables are also infested heavily by several major insect pests.

Brinjal *Solanum melongena* L.

Fruit and shoot borer *Leucinodes orbonalis* Gunnee (Lepidoptera: Pyraustidae)

Economic importance: Most destructive pest of brinjal, found throughout the country. Besides brinjal, the pest is also known to infest potato, bitter gourd, pea pods, cucurbits and other solanaceous etc. The infestation on brinjal can be as high as 70 per cent. The pest is active throughout the year, except in severe cold weather in North India.

Mark of identification and life cycle: Moths wings are white with pinkish or black tings and are ringed with small hair along the apical and anal margins. The fore wings are ornamented with a number of black, pale and light brown spots. Female lays 80-120 creamy white eggs, singly or in batches of 2-4 on the underside of leaves, on green stems, flower buds or the calyces of fruits. Caterpillars are creamy when young, but light pink when full-grown in 9-28 days. Pupate in tough silken cocoons among the fallen leaves with period of 6-17days and adult survive for 2-5days of its life cycle. A single life-cycle takes 22 - 25 days extending up to 74 days during winter and there may be 8 - 12 generations in a year.

Nature of damage: Infestation starts after few weeks of transplantation. The larva bore into the growing shoots or petioles of large leaves and feed on internal tissues. As a result of damage, affected shoots wither and plants exhibit the symptoms of drooping. After fruit formation larva makes their entry under the calyx when they are young. The holes later plugged with excreta leaving no visible sign of infestation. Large holes seen on the fruits are the exit holes.

Carry over: The remain of brinjal plant stalks from previous crops as most farmers store these plants around their fields and use the dried stalks as fuel for cooking.

Management strategies

- Cultivation of fruit shoot borer tolerant varieties such as Punjab Barsati, Punjab neelam and BH-2.
- Continuous cropping of brinjal and potato should be avoided and resistant varieties if available should be cultivated.
- Removal and destruction of affected shoot and fruits along with larvae.
- Installation of pheromone traps for monitoring and mass trapping or use light traps @ 1/ha to attract and kill insects.
- Brinjal with long, narrow fruits are less susceptible to attack and, therefore, this variety should be preferred than other varieties.
- Inter cropping with single and double rows of coriander *Coriandrum sativum* (L.) as well as border crop of brinjal significantly reduced the incidence of brinjal shoot and fruit borer.
- Clipping of infested shoots contributed less damage.
- Synthetic pyrethroids in hot weather are less effective. Avoid using synthetic pyrethroids as they also cause resurgence of sucking pests.
- **Natural enemies**: *Pristomerus testaceus* Mori and *Cremastus flavorbitalis* Cam. (Ichneumonidae). *Eriborus argentiopilosus Bracon* sp., *Shirakia shoenobii* Vier and *Iphiaulax* sp. (Braconidae), *Apanteles* spp., *Chelonus* spp, and the chalcids *Brachymeria obscurata* (Walker).
- Application of neem cake four times during the crop growth decreased the incidence of borer to eight per cent as against 40 per cent in control.
- Spray *B. thuringiensis* var kurstaki @ 1.5 and 2 ml / lit of water.
- Spraying once or twice at fortnightly interval with 200ml Ripcord 10 EC (cypermethrin), 160ml Decis 2.8 EC (deltamethrin), 100ml Sumicidine 20 EC (fenvalrate), 800ml Ekalux 25EC (quinalphos), 500ml Hostathion 40EC (triazophos).
- Chlorantraniliprole (40-60g a.i /ha) (Coragen 20SC) is highly effective (Rajavel *et al*., 2011).
- Avoid using insecticide at the time of fruit maturation and harvest.

Brinjal fruit and shoot borer, infested twig, eggs, larvae and adults with damaged fruit

Brinjal Stem Borer, *Euzophera perticella* Rag. (Phycitidae: Lepidoptera)

Distribution: It is found in different parts of South East Asia. The insect is considered to be one of the important pests of brinjal, sometimes becoming serious on it. It also breeds on brinjal, other solanaceous plants like *Solanum tuberosum,* S. *nigrum, S. xanthocarpum. S. torvum,* tomato, *Datura* sp., *Physalis* sp. and *Withania somnifera.*

Nature of damage: The grubs and adults scrape the leaves in a characteristic manner and feed. They feed on the epidermal layers of leaves which get skeletonized and gradually dry away. They affect the crop in all the stages.

Life history: The brownish hemispherical beetle has 12-28 black spots on the elytra. The female lays elongate, spindle-shaped yellowish eggs in groups of 10 - 20 on the under surface of leaves. About 120-180 eggs may be laid by a female. The egg period is 2-4 days. The yellowish spiny grubs become full grown in 10 -35 days and pupate on the leaf or stem. The pupa is hemispherical, yellowish with spines on the posterior part. The anterior portion being devoid of spines. Adults emerge in a week and live for a month feeding on leaves. The total life-history takes 17 - 50 days depending on weather conditions. (Atwal, 2005)

Management practices/strategy

(i) In the initial stage, collection and destruction of affected leaves alongwith the eggs, grubs and adults.

(ii) Spray NSKE 5 per cent in initial stage of crop.

(iii) If still the infestation persist spray Emamectin benzoate 1.9% EC @ 1.5ml/l or Indoxacarb 14.5% SC/Avant @1.0ml/l or Qinalphos @ 1.0ml/l or Malathion @1.5ml/l or Spinosad 2.5%SC @ 1ml/l.

(iv) Waiting period must be followed after each spray.

Grey weevil, *Myllocerus subfasciatus* G. or *M. maculosus* (Curculionidae: Coleoptera)

Distribution: It is a polyphagous pest occurring on a number of crops like cotton, sorghum, pearl millet and maize all over India.

Nature of damage: The adult beetles feed on leaves of brinjal and the grubs feed on roots and cause wilting and death of plants. Occasionally the insect assumes serious proportions on the crop.

Life history: The brownish weevil lays about 500 eggs in the soil about 80 - 100mm deep. The incubation period is 3-11 days. The grubs become full grown in 28 - 34 days and pupate in the soil-in earthen cocoons. The adults emerge in about 5 - 7 days. Total life cycle is completed in 6-8 weeks.

Management strategies

(i) Inter-culture of the crop regularly to prevent population build up and carryover of these weevils.

(ii) Spray NSKE 5 per cent in initial stage of crop.

(iii) If still the infestation persist spray Emamectin benzoate 1.9 %EC @ 1.5ml/l or Indoxacarb 14.5 %SC/Avant @ 1.0ml/l or Qinalphos @ 1.0ml/l or Malathion @1.5ml/l or Spinosad 2.5%SC @ 1ml/l.

(iv) Waiting period must be followed after each spray (Table1).

Leaf hopper, *Amrasca biguttula biguttula* (= *Empoasca devastans* Dist.) (Cicadellidae: Homoptera)

Distribution: Distributed all over India; but particularly serious in Sind, Punjab and Tamil Nadu. It is a polyphagous pest attacking okra, brinjal, beans, castor, cucurbits, hollyhock, potato, sunflower and other malvaceous plants.

Nature of damage: The nymphs and adults remain on the under surface of the leaves and suck the cell sap and while feeding inject their toxic saliva. As a result the plant become stunted, the leaves crinkle, turn yellowish and become cup shaped. Brownish or reddish colour may develop along the edges of the leaves. This is called the 'hopper burn'. This pest is very active from September to January.

Life history: The adult is a slender green insect. Elongate, yellowish eggs are laid singly inside the leaf vein on the under surface of the leaves. A female lays 15-30 eggs, leaves of 35-40 days old are preferred for egg laying. Egg hatch in 4-10 days. Nymphs are green and wedge shaped. The first and second instar nymphs feed mostly near the base of the leaf vein. Then they distribute themselves throughout the leaf and feed from the under surface. The nymphs develop into adults in 7-21 days. They breed throughout the year.

Management strategies

(i) Clean cultivation is effective tool for reducing pests population.

(ii) Conserve the spider population in the field.

(iii) Grow other group of crop like cabbage, okra etc. in next season.

(iv) If still the aphid population persists spray Imidacloprid 70%WG @ 0.5ml/l or Acetamiprid 20 %SP @ 0.2g/l or Thiamethoxam 25 %WG@ 0.2g/l or Metasystox @ 1.5ml/l.

(v) Waiting period must be followed after each spray (Table1).

Hadda beetles *Epilachna* spp. (Coleptera : Coccinellidae)

Economic importance: Commonly occurs throughout south-east Asia. Cause severe damage during May to September. It also attacks bitter gourd, bottle gourd, potato and tomato.

Mark of identification and life cycle: Beetles of all the three species are about 8-9 mm in length and 5-6 mm in width. *E.viginticotopunctata* beetles are deep red and usually have 7-14 black spots on each elytron whose tip is somewhat pointed. Beetles of *E.dodecasitigma* are deep copper-coloured and have six black spots on each elytron whose tip is more rounded. *E. demurili* beetles have a dull appearance and are light copper-colored. Each of their elytron bears six spots surrounded by yellowish rings. Adult lays as many as 120 to 180 yellow cigar-shaped eggs, mostly on the underside of leaves, in batches of 5-40 each. Grubs of all the three species are about 6 mm long, yellowish in colour and have six rows of long branched spines and become full grown in 10-35 days. The pupae are darker and are found fixed on the leaves stems and most commonly at the base of the plants. It passes the winter as hibernating adults among heaps of dry plants or in cracks and crevices in the soil. Life cycle is completed in 25-50 days.

Nature of damage: Both grubs and adults feed by scraping chlorophyll from epidermal layers of leaves, leaving the veins and veinlets, and cause characteristic skeletonized patches on the leaves and forming ladder-like windows. In severe cases even calyx of the fruit may also be infested. Later, the affected areas on leaves dry and fall off and damage appears in the form of holes in the leaves. Infested leaves turn brown dry up and fall off and completely skeletonize the plants.

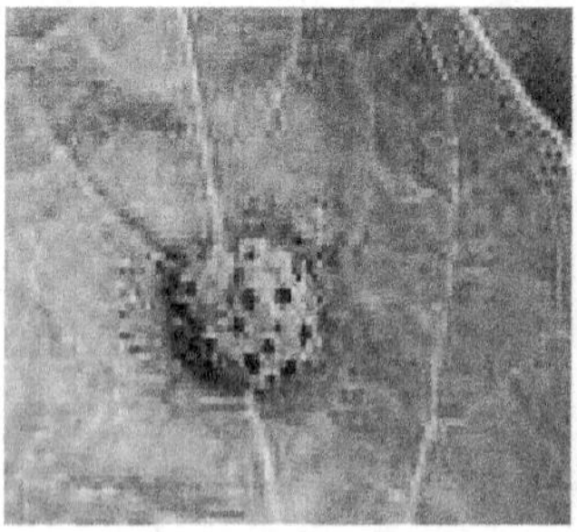

Adult

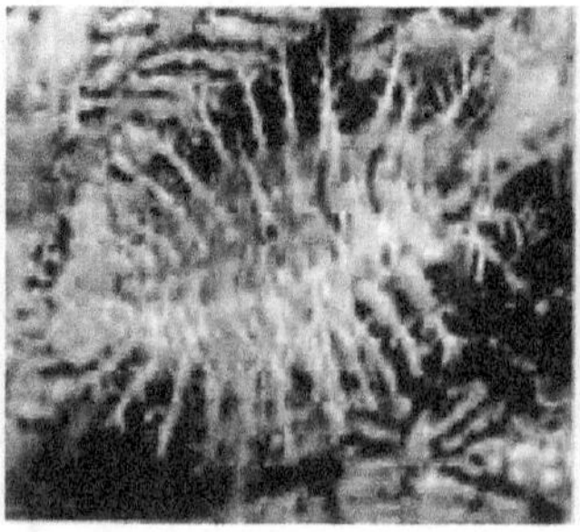

Grub

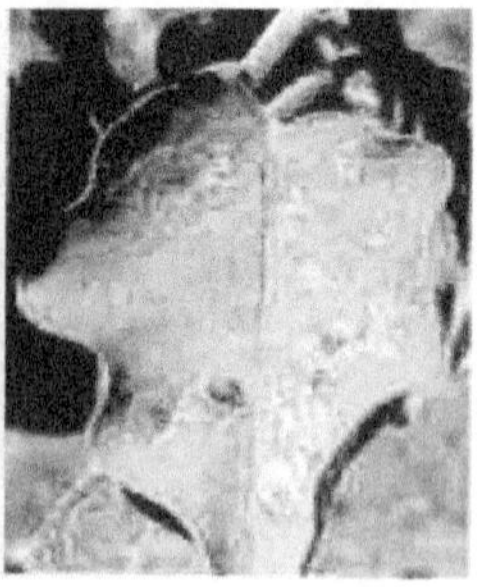

Damaged Leaf

Management strategies

- Collect and destroy egg masses and skeletonised leaves with adults and grubs.
- Hand picking of grubs and collection of beetles by hand nets in the early stages of attack is recommended for small holdings.
- **Natural enemies**: *Tetrasatichus ovulorum* Ferr., *Chrysonotomia appannai* L. and *Pediobius foveolatus*.
- On appearance of beetle spray 250ml malathion 50EC 100-125litres of water / acre at 15 days interval.
- Mix diflubenzuron invariably with chlorpyriphos 0.05% and spray on the crop which reduces the population by nearly 95% in field.

Jassid

The jassid acts as a vector of little-leaf of brinjal from diseased to healthy eggplants caused by a virus.

White fly *Bemisia tabaci* Gennadius (Homoptera: Aleyrodidae)

Economic importance: Found throughout India. Polyphagous in nature with its main hosts as cotton, tobacco, and some winter vegetables. Cotton whitefly, which is a vector of leaf curl virus has been observed in increasing severe form on tomato and brinjal and other vegetable crops.

Mark of identification and life cycle: same as chilli

Nature of damage: On hatching nymphs crawl a little, settle down on a succulent spot on ventral surface of leaf, and keep sucking sap. Affected parts become yellowish, leaves wrinkle, curl downwards and are ultimately shed. Honeydew excreted by nymphs attracts sooty molds which form black coating on leaves.

Whiteflies

Damaged leaf

Management strategies

- As the flies are attracted to yellow colour, place yellow plates / tins with grease so that the attracted flies get stuck to the grease.
- To deal with lower levels, place yellow sticky traps to monitor and suppress infestations.
- Release **natural predators** such as ladybugs, lacewings, or whitefly parasites.
- If 5-10 whiteflies are noticed per leaf, spray Triazophos 2.5 ml or Profenophos 2.0 ml/litre of water.
- Do not spray synthetic pyrethroids like Deltamethrin, Cypermethrin and Fenvalerate as it leads to resurgence.

Brinjal Lace-wing Bug *Urentius sentis* Distant (Hemiptera: Tingidae)

Economic importance: This pest is found in plains of India and is not a serious pest. Brinjal is the only host.

Mark of identification and life cycle: Bugs measure about 3 mm in length and are straw coloured on the dorsal side and black on the ventral side. On the pronotum and wings, there is a network of markings and veins. Adult lay 35-44 shinning white nipple-shaped eggs singly in the tissues on the underside of the leaves. The eggs hatch in 3-12 days. Nymph are pale ochraceous and are stoutly and built, with very prominent spines with a life period of 10-12 days. The adult overwinter from November to March in cracks in the soil under brinjal plants and start ovipositing in April.

Nature of damage: The adults and the nymphs suck the sap form leaves and cause damage by yellowish spots which, together with the black scale-like excreta deposited by them, impart a characteristic mottled appearance to the infested leaves. The young nymphs feed gregariously on the lower surface of the leaves, but the fully-developed nymphs are found feeding and moving about individually on the lower surface as well as on the upper surface.

Management strategies

- Collect and destroy egg masses and skeletonised leaves with adults and grubs.
- Hand picking of grubs and collection of beetles by hand nets in the early stages of attack is recommended for smallholdings.
- Release of lady beetle larvae and adults, spiders and shield shaped soldier bugs.
- If infestation is severe 250ml malathion 50EC in 100-125litre water/acre at 10 days interval.

Red spider mite *Tetranychus cinnabarinus* Boisduval (Tetranychidae: Arachnida)

Economic importance: Highly polyphagous and has a world-wide distribution. Mite attack during April-June and are very serious when the conditions are hot and dry. It feed on more than 150 host plants including vegetables like brinjal, ladies finger, cowpea, cucurbits, beans etc.

Mark of identification and life cycle: Adults are ovate, reddish brown with four pairs of legs. Eggs are globular and whitish. Larvae (I^{st} *instar* nymphs) are pinkish with three pairs of legs while nymphs (later instars) are greenish-red, look like the larvae, but have four pairs of legs. Life cycle is completed in about 20 days.

Nature of damage: Colonies of mites are found feeding on ventral surface of leaves under protective cover of fine silken webs, resulting in yellow spots on dorsal surface of leaves. Affected leaves gradually curl, get wrinkled and crumpled. Mite attacked leaves attract lot of dust particles. In heavy infestations even fruits are affected.

Management strategies

- Avoid ratoon crop of brinjal.
- Adequate irrigation is important because water stress plants are most likely to be damaged.
- Sprays of water, insecticidal oil, or soaps can be used.
- Use synthetic insecticides when required
- 300ml Omite 57 EC/Kelthane 18.5 EC or 450ml Phosmite 50EC or 250ml Metasystox 25EC in 100 to 150 liters of water per acre.

Tomato *Lycopersicon esculenta* Milliere

Tomato crop is cultivated over 0.47 m hectares in India and biotic stress causes 30-35% loss in crop yield (Sardana and Sabir, 2007).

Fruit borer *Helicoverpa armigera* Hubner (Lepidoptera: Noctuidae)

Economic importance: Tomato fruit borer is the major limiting factor in the successful cultivation of this crop. It has caused 50-80 per cent fruit damage in different years (Devi *et al*., 2015).

Mark of identification and life cycle: Adults are pale brown yellow nocturnal moths with wingspan of 40 mm. Forewings are provided with series of black dots. The undersurface of forewings carry a black kidney shaped mark. The hindwings are yellowish brown with broad black outer band.Egg are shining green, white ribbed and dome shaped eggs (300-600 eggs per female) laid singly on tender parts of the plant. The stoutly built caterpillar of variable colour possesses prominent noticeable dark brown broken lines along the body. The full-grown caterpillar is about 40 mm long. Pupa is dark brown in colour and have a sharp spine at the posterior end.

Nature of damage: The young larvae feed on the foliage for some time and later bore into the pods and feed on the developing grains, with their bodies hanging outside. Although they prefer food plants like gram and red gram, the larvae are polyphagous. They feed on the foliage, when young and on the seed in later stages, and thus reduce yield. A single larva may destroy 30-40 pods before it reaches maturity.

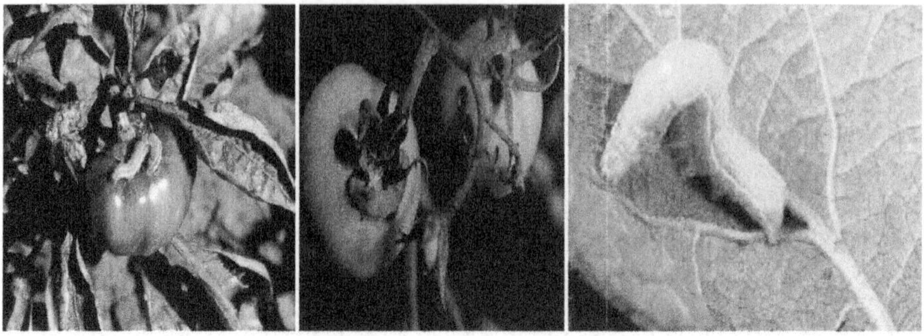

Tomato fruit borer Damage Semilooper

Management strategies

- Grow less susceptible genotypes Rupali, Roma, Pusa red plum.
- Hand picking of larvae: Larvae of cutworm, leaf eating caterpillar are very sluggish, so they can be hand collected and destroyed easily.
- Burning: Damaged fruits and crop residue should be burn to avoid carryover of pest or cold storage of fruits and vegetables reduces pest infection.
- Setup pheromone trap with Helilure at 5/ha and change the lure once in 15 days.
- Light traps are used for many pests like hairy caterpillar, stem borer.
- Ploughing the field after summer showers, removing the crop debris from the field, exposing the different stages of insects viz., egg, larvae and pupae to sunlight greatly reduce the pest abundance and prevent the pest population buildup.
- Planting of yellow tall marigold (*Tagetes Spp.*) or *Bidil rustica* tobacco around tomato (1:5) has been found promising. All the eggs of *H.armigera* deposited on yellow Targets flower buds could be destroyed by the inundation of *Helicoverpa* adapted strain of egg parasitoid (*Trichogramma Chilonis*).
- Six releases of *T. chilonis* @ 50,000/ha per week coinciding with flowering time and based on ETL.
- Release *Chrysoperla carnea* at weekly interval at 50,000 eggs or grubs / ha from 30 days after planting.
- Encourage activity of parasitoid *Eucelatoria bryani, Campoletes, Chelonus* etc.

- Spray HaNPV at 500 LE/ha along with cotton seed oil 300 g/ha to kill larvae.
- *B. thuringiensis* 2 g/lit or quinalphos 2.5 ml/lit.
- Careful destruction of damaged and disease affected tomato fruits after harvesting.
- Neem seed kernel extract (2 to 5%) has been found effective against several pests including cutworm, plant hopper, leafhoppers, tobacco caterpillar, several species of aphids and mites.
- Bird perches @ 10 /acre should be install for facilitating field visits of predatory birds.
- Spray imidacloprid or thiomethoxam at 15 days interval for leaf miner and whitefly control.
- If borer cross ETL level apply 200ml indoxacarb 14.5 SC, fame 480 SL (fludendamide) 30 ml, carina 50 EC (profenophos) 600 ml, ripcord 10 EC (cypermethrin) 200ml, decis 2.8 EC (deltamethrin) 160 ml per acre or spinosad 45 SC 73a.i./ha (Ghosh *et.al.*, 2010)
- Do not spray insecticides after maturity of fruits.

Whitefly – *Bemisia tabaci* (Aleyrodidae: Hemiptera)

See chilli whitefly

Management strategies

(i) Spraying with 0.05% formothion or dimethoate or 0.01% imidacloprid or 0.1% acetamiprid.

(ii) In case of severe infestation, two sprayings at 10 - 12 days interval with 0.03% oxydemeton methyl or thiometon.

Leafminers *Liriomyza* spp.

Economic importance: It can reach damaging levels quite rapidly if certain disruptive insecticides are used repeatedly. *Liriomyza trifolii*, has been the most common leafminer pest of tomatoes since 1990. There has been a recent change in the pest status of a related species. Reports are incomplete at this time on its status as a pest on tomatoes in India, but other parts of the world report significant losses on fresh market tomatoes.

Mark of identification and life cycle: Leafminer adults are small, black and yellow flies. Eggs are inserted in leaves and larvae feed between leaf surfaces, creating a meandering track or "mine." At high population levels, entire leaves may be covered with mines. Mature larvae leave the mines, dropping to the ground to pupate. The life cycle takes only 2 weeks in warm weather; there are seven to ten generations a year. All three species feed on a wide variety of crops and weeds; development continues all year and the population moves from one host to another as new host plants become available.

Nature of damage: Leafminer feeding results in serpentine mines (slender, white, winding trails); heavily mined leaflets have large whitish blotches. Leaves injured by leafminers drop prematurely; heavily infested plants may lose most of their leaves. f it occurs early in the fruiting period, defoliation can reduce yield and fruit size and expose fruit to sunburn. Pole tomatoes, which have a long fruiting period, are more vulnerable than other tomato crops. Leafminers are normally a pest of late summer tomatoes and can reach high numbers.

Management strategies

- Check transplants before planting and destroy any that are infested
- Several species of parasitic wasps, particularly *Chrysocharis parksi* and *Diglyphus begini* attack leafminer larvae; left undisturbed, parasites often keep leafminers under control.
- Where a series of tomato crops is planted in the same area, reduce early infestations in a new crop by removing old plantings immediately after the last harvest.
- Biological and cultural controls as well as sprays of the Entrust formulation of spinosad are acceptable for use on an organically certified crop.
- Treatment recommendations currently involve the rotation of abamectin and cyromazine. Some species are also controlled to a certain degree by spinosad.

White Tailed Mealy Bug, *Ferrisia virgata* (Cockerell) (Coccidae: Homoptera)

Distribution: It is pan - tropical in distribution and is found all over the Indian subcontinent and South-East Asia. It is polyphagous and has a very wide range of host plants including, beans, cashew, cassava, coffee, cocoa, citrus, cotton, groundnut, guava, jute, sugarcane, sweet potato and tomato. It is found throughout the year, though it is less active during winter.

Nature of damage: Eggs are laid in clusters in cottony ovisac which remains concealed under the female. On hatching, the crawlers remain huddled together in cottony nest under the body of the mother. Later, these crawlers become active and wander about, moving swiftly till they find a succulent spot where they puncture the epidermis, inject their toxic saliva and start sucking the cell sap. The mechanical injury thus caused also serves as an entry for various disease producing organisms (bacteria and fungi). From 2nd instar onwards the nymphs secrete honeydew on which black sooty mould develop, which in turn hinders the photosynthetic activity of the plant resulting in stunted growth.

Life history: Eggs of this mealy bug are pale-yellow, cylindrical and about 0.3 mm long. A single female lays 100-400 eggs which remain concealed under the female. Freshly hatched crawlers are yellowish in colour and become pale white in 2-3 days. Adult females are apterous, long, slender, slightly oval (3.5 – 4.5 x 1.5 – 2.0 mm) covered with dusty white waxy secretion and having a pair of conspicuous long glossy wax tassels at the caudal end. Reproduction is sexual as well as parthenogenetic. Incubation period is 15 minutes to 4 hours and the immature stages may last for about 20 - 60 days in case of male and 19 - 47 days in case of females. Longevity of males is 1 - 3 days while the females live for 5–7 weeks.

Management strategies

(i) Remove and destroy mechanically all the affected leaves and twigs in the early stages of infestation.

(ii) Use Imidacloprid 17.8 SL @ 0.5ml/l or Acetamiprid 20 %SP @ 100g/ha or Thiamethoxam 25 %WG @ 100g/ha within one month of transplanting.

Nematodes: This is one of the most dreaded tomato problems especially in poly house that are growing tomato. Actually, almost 20,000 different species of nematode have been identified, and billions of these usually microscopic worms occupy each acre of fertile earth, so it is fortunate that only a few cause gardening problems. This particular species invades various crops, causing bumps or galls that interfere with the plant's ability to take up nutrients and to perform photosynthesis. They're most common in warmer areas with short winters. Unfortunately, controlling nematodes is not easy.

Management strategies

- Rotation: Since they take several seasons to get established, rotation with non host crops denies this pest the chance to get entrenched.
- Soil sterilization: Completely sterilizing the soil is one option on small plots, but it's toxic and sometimes expensive. It also means that you've killed off all the beneficial organisms in the soil as well as the troublesome ones, so it's particularly important to follow such treatment with a big infusion of clean compost. It would also be best to add earthworms, and an assortment of micro-organisms, since doing so will restore the soil to full health and make it less vulnerable to further incursions by nematodes.

- While eliminating nematodes is extremely difficult, it is possible to limit their damage by using resistant varieties, Grow nematode resistant variety Punjab NR-7 in infested fields.
- Incorporate 40 days old Toria and Taramira crops into tomato nursery beds 10 days before sowing and turn the soil 3-4 times before sowing of tomato.
- Dip the the roots of nursery plants in 0.03% dimethoate (10ml Rogor 30 EC in 10liters of water) for 6 hours before transplanting.
- Grow garlic in root knot nematode infested fields in rotation with other vegetable crops (Annonymous, 2015).

Chilli *Capsicum annum* L.

Chilli is an important spice and vegetable crop of India and is grown for both market and export, contains capsaicin which has a medicinal value. It has an important nutritive value especially rich in vitamin C and A. About 51 species of insects and 2 species of mites belonging to 27 families under 9 orders along with 2 species of snail and two species of millipedes are know to damages chilli crop both in nursery and in the field (Rahman *et al.*, 2011). Among these pests fruit borers, thrips, mites are of serious in nature.

Cotton whitefly *Bemisia tabaci* Gennadius (Hemiptera: Aleyrodidae)

Economic importance: Commonly known as cotton whitefly is found in most of the countries in tropics and subtropics. Its main hosts are cotton, tobacco and some winter vegetables; including tomato, the infestation on these crops is sporadically severe.

Mark of identification and life cycle: Adult is yellowish bodies are slightly dusted with a white waxy powder. They have two pairs of pure white wings and have prominent long hind wings. The females lay eggs singly on the underside of the leaves. The eggs are stalked, sub-elliptical and light yellow at first, turning brown later on. Incubation period is 3 - 5 days in summer extending up to 33 days during winter. The nymphs are elliptical and soon fix their mouthparts into the plant tissues. Nymphal development takes 9 - 14 and 17 - 81 days in summer and winter, respectively and pupal period lasts for 2 - 8 days being longer during winter than in summer. A life-cycle may be completed in as little as 14 days or it may even be prolonged up to 107 days. There are about 12 overlapping generations in a year.

Nature of damage: White, tiny, scale-like insects may be seen darting about near the plants or crowding in between the veins on ventral surface of leaves, sucking the sap from the infested parts. The pest is more active during the dry season and its activity decreases with the onset of rains. As a result of their feeding the affected parts become yellowish, the leaves wrinkle and curl downwards and are ultimately shed. Besides the feeding damage, these insects also exude honeydew which favours the development of sooty mould. From a distance, the attacked crop gives a sickly, black appearance. Consequently, the growth of the plants is adversely affected and when the attack appears late in the season, the yield is lowered considerable. *B. tabaci* is known to transmit a number of virus diseases including the leaf curl disease of tobacco, the vein clearing disease of okra and the leaf-curl of sesame.

Chilli Thrips *Scirtothrips dorsalis* Hood (Thysanoptera: Thripidae)

Economic importance: *Scirtothrips dorsalis* is a polyphagous species and has been documented to attack more than 100 recorded hosts from about 40 different families. As this pest expands its geographical range additional plants are added to its host range.

Mark of identification and life cycle: Female thrip lays single kidney shape egg inside the tissues of the leaves and shoots. Reproduction is both sexual and parthenogenetic. In case of sexual reproduction, oviposition period lasts for about a month during which a female lays on an average 100 eggs @ 2-4 eggs per day. The nymphs resemble the adults in shape and colour but are wingless and smaller in size. Adult is slender, yellowish brown in colour, having apically pointed wings, and they measure about 1 mm in length. The females possess long, narrow wings with the fore margin fringed with long hairs. A single life-cycle is completed in 2-2½ weeks as many as 25 overlapping generations in a year.

Nature of damage: Damage is caused by the adults as well as by the nymphs. They suck the cell sap from tender regions and cause the leaves to shrivel. The attacked plants are stunted and may finally dry up. The insect is also responsible for transmitting the virus causing leaf curl disease of chillies. In case of severe infestation, there is malformation of leaves, buds and fruits, which may damage half the crop.

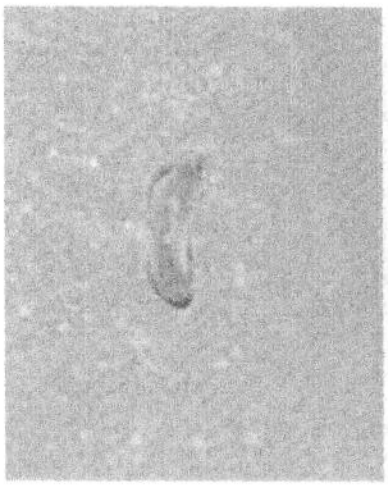
Egg

Nymph

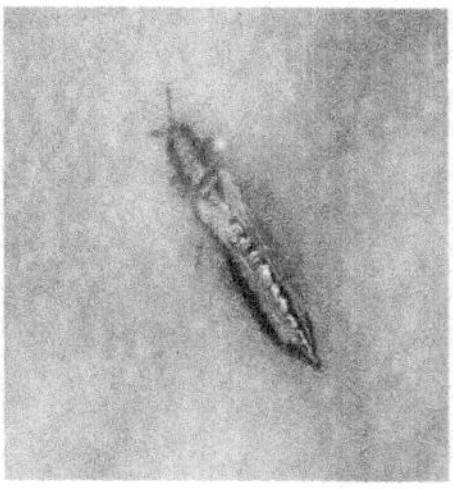
Adult

Damaged leaf

Chilli mite *Polyphagotarsonemus latus* (Banks)

Economic importance: Mites or acarids are tiny, spider-like creatures.

Mark of identification and life cycle: The adults are eight legged while the larvae are six legged. The body shape is round and sac-like and un-segmented. They are slightly yellow in colour, very small in size (0.1–0.2 mm long) and very mobile. They are easily dispersed by wind and their populations normally build up during the dry weather.

Nature of damage: Both adults and larvae suck sap from the leaves resulting in the leaves curling downwards. In severe infestation, scarring of the stem and fruit skin is also common. Frequent watering of plants during the dry weather helps to reduce the pest population.

Chilli mite damaged plant

Chilli fruit borer *Helicoverpa armigera* Hubner and *Spodoptera litura* Fb.

Economic importance: This pest is gaining importance in cultivation of chilli in North India. Now became an important pest for chilli growers. The damage caused by *H. armigera* and *S. litura* during flowering and fruit formation is the most concern. As reported by Reddy and Reddy (1999) due to severe attack of fruit borers lead to 90 per cent flower and fruit drop in chilli.

Mark of identification and life cycle: same as in tomato (*H. armigera*) and cauliflower (*S. litura*).

Nature of damage: These two species bore into the tender and maturing fruits and feed the seed inside resulting in hallowing of the fruit and finally dropping of the infested fruit from the plant.

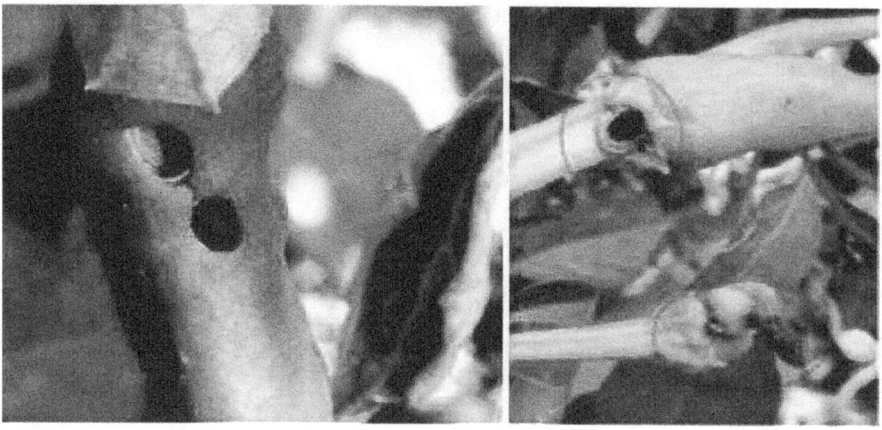

Damaged chilli fruit

Management strategies

- Changing the date of sowing and transplanting reduce the attack of sucking pest. Sowing of nursery on 15th June and transplanting on 15th July recorded significantly low sucking pest population under south Indian rainfed conditions (Gayatridevi and Giraddi 2009).
- Seed treatment with Imidacloprid (Gaucho) @ 5 grams per kg seed for the management of sucking pest.
- Spray Imidacloprid @ 1 ml in 3-4 liters of water or fipronil @ 2 ml per liter or 400 ml malathion 50 EC in 100-125 liters water at 15-20 days interval per acre.
- Chilli crop bordered by two rows of maize at every 0.5 acre area with two interventions of spray with neemazal 1% @ 2ml per liter at 7 week after transplanting and second spray with difenthiuron 50 WP @ 0.75 g per liter at 9 week after transplanting recorded least leaf curl damage due to thrips and mite (Tatagar *et al.*, 2011)

- Marigold trap crop, vermicompost 2.5 t/ha + Neem cake 250kg/ha (without application of recommended dose of fertilizers, i.e RDF) superimposed with sprays of Neemazal @ 2ml/l at 5 week after transplanting(WAT), Diafenthiuron @ 1g/l (8 WAT), profenofos@ 2 ml/l(11 WAT) and Neemazal@2ml/l(14 WAT) (Gundannavar et al., 2007).
- Thrips predators- minute pirate bugs and green lace wing (grub and adult) (chilli thrips), *Eretmocerus massii* Silv. (Aphelinidae), *Encarsia* sp., *Chrysoperla* sp. (Chrysopidae) and *Brumus* sp. (Coccinellidae) (whitefly). The thrips *Franklinothrips vespiformis* (Crawford) and *Erythrothrips asiaticus* R. &. M. are also predaceous on this thrips
- Flubendiamide 20 WG @ 60 g a.i. /ha against chilli fruit borers, emamectin benzoate 5 SG @ 11 g a.i./ha (7.22 q/ha) and spinosad 45 SC @ 75 g a.i./ ha (7.32q/ha). (Tatagar *et al.*, 2009) or Neem leaf extract @ 0.5kg/2litre of water at 7 days interval (Rahman *et al.*, 2011).
- Spray with miticides such as dicofol @ 5 ml per liter or wettable sulphur 3 grams per liter or Pegasis @ 1 gm per liter or Vertemic @ 0.5 ml per liter. Use overhead irrigation with sprinklers for effective management of mites wherever possible. (ICRISAT. 2011)

White grubs *Holotrichia* spp. (Coleoptera: Scarabaeidae)

Mark of identification and life cycle: The adult beetles lay eggs singly up to a depth of 10 cm. Grub of ber beetle is white, having a brown head and prominent thoracic legs. The grubs are mostly found in the upper 5-10 cm layer of soil. The shape of a pupa is semicircular and creamy white. The adult beetles are dull brown and measure about 18 mm in length and 7 mm in width.

Nature of damage: The grubs eat away the nodules, the fine rootlets and may also girdle the main root, ultimately killing the plants. At night, the beetles feed on foliage and may completely defoliate even trees like neem (*Azadirachta indica*) and banyan (*Ficus bengalensis*).

Natural enemies: *Scolia aureipennis* (Scoliidae). A fungus, *Metarrhizium anisopliae* (Metchnikoff) (Moniliaceae) parasitizes the adults and the common Indian toad, *Bufo melanostictus*, and the wall lizard, *Gecko gecko* feed on the beetles.

References

Annonymous 2015. Package of practice for cultivation of vegetables, Punjab Agricultural University, Ludhiana.

Anonymous, 2015. www.plantnatural.com/tomato-gardening guru. 6.7.15.

Devi, L.L., Senapati,A.K. and Chatterjee, M.L. 2015. Biorational management of tomato fruit borer, Helicoverpa armigera at new alluvial zone of west Bengal. In 4th congress on insect science "Entomology for sustainable Agriculture .April 16-17 at PAU, Ludhiana

Dhandapani, N., Umeshchandra, Shelkar, R. and Murugan, M. 2003. Bio-intensive pest management (BIPM) in major vegetable crops: An Indian perspective. Food, Agriculture & Environment Vol.1 (2): 333-339.

Gayatridevi, S. and Giraddi, R.S. 2009. Effect of date of planting on the activity of thrips, mites and development of leaf curl in chilli (*Capsicum annum* L.) *Karnataka Journal of Agricultural Sciences* 22 (1):206-207.

Ghosh,A., Chatterjee, M. and Roy, A. 2010. Bio-efficacy of spinosad against tomato fruit borer (*Helicoverpa armigera* Hub.) (Lepidoptera: Noctuidae) and its natural enemies. *Journal of Horticulture and Forestry* 2(5):108-111.

Gundannavar, K.P., Giraddi, Kulkarni, K.A. and Awaknavar, J.S. 2007. Development of integrated pest management modules for chilli pests. *Karnataka Journal of Agricultural Sciences* 20 (4):757-760.

ICRISAT. 2011. Integrated Pest Management in Chilli, Tomato, Onion and Potato crops.PP-4.

Krishnamoorti, A. and Mani, M. 1988. Feasibility of managing *H. armigera* on tomato through parasitoid. In: National Workshop on *Heliothis* management. Feb. 18-19. TNAU, Coimbatore.

Rahman, M.M., Rahman, S.M.M. and Akter, A. 2011. Comparative performance of some insecticides and botanicals against chilli fruit borer. *Journal of Experimental Sciences* 2(1): 27-31.

Rai, A.B., Loganathan, M. Halder, J., Venkataravanappa, V. and Naik, P.S. 2014. Eco-friendly Approaches for Sustainable Management of Vegetable Pests. IIVR Technical Bulletin No. 53, IIVR, Varanasi, pp. 104.

Rajavel, D. S., Mohanraj, A. and Bharathi, K. 2011.Efficacy of chlorantraniliprole (Coragen 20SC) against brinjal shoot and fruit borer, *Leucinodes orbonalis* (Guen.). *Pest Management in Horticultural Ecosystems*, 17(1): 28-31.

Reddy, M.R.S. and Reddy, G.S. 1999. An eco-friendly method to combat *Helicoverpa armigera* (Hub.). *Insect Environ.* 4: 143-144.

Sardana, H.R. and Sabir,N. 2007. IPM strategies for tomato and cabbage. Extension folder by NCIPM, New Delhi.

Sardana, H.R., Tyagi, A. and Singh, A. 2005. Knowledge resources on fruit flies (Tephritidae: Diptera) in India. National center for integrated pest management, New Delhi, India.

Tatagar M. H., Mohankumar H. D., Shivaprasad M. and Mesta R. K. 2009. Bio-efficacy of flubendiamide 20 WG against chilli fruit borers, *Helicoverpa armigera* (Hub.) and *Spodoptera litura* (Fb.) *Karnataka J. Agric. Sci.* 22(3-Spl. Issue): 579-581.

Tatagar, M.H, Awaknavar, J.S., Giraddi, R.S., Mohankumar, H.D., Mallapur, C.P. and Kataraki, P.A. 2011. Role of border crop for the management of chilli leaf curl caused due to thrips, *Scirtothrips dorsalis* (Hood) and mites, *Polyphagotarsonemus latus* (Banks). *Karnataka Journal of Agricultural Sciences* 24 (3):294-299.

8

Fruit Crops Insect Pests and Their Biointensive Integrated Pest Management Techniques

Reena and Bhav Kumar Sinha

A systems approach to pest management based on an understanding of pest ecology is referred to as Biointensive Integrated Pest Management (BIPM). It is more dynamic and ecologically informed as compared to IPM that considers the farm as part of an agro ecosystem. Accurate diagnosis of the pest problem is the first step, which then relies on a range of preventive tactics and biological controls to keep pest populations within acceptable limits. It also considers ecological and economic factors into agricultural system design and decision - making and addresses public concern about environmental quality and food safety. As a last resort, the reduced – risk pesticides are used if other tactics have not been adequately effective.

The earlier concept of IPM has failed because of ease of applicability to easy availability of chemical pesticides. Hence, the present concept of BIPM, with more intensive use of biological and other biorational means has emerged. The flexibility and environmental compatibility of BIPM strategies make it useful in all types of cropping system. BIPM would likely decrease chemical use and costs entailing them. BIPM ultimately provides resistance / immunity to the entire system so that no foreign elements (insect pests or disease causing organisms) survive or gain upper hand. BIPM if followed for years shall enhance the natural control system, thus suppressing pest populations below those causing economic injury.

BIPM shares almost all the components of conventional IPM including monitoring, use of economic thresholds, record keeping, planning and all the methods of control except chemical means. The main emphasis of BIPM is on proactive measures to redesign the agricultural ecosystem to the disadvantage of insect pest and to the advantage of its parasites, parasitoids and predators complex. Good planning is must before the implementation of any IPM program and is more important in BIPM. It should be done well in advance, taking beneficial organisms' habitat manipulation, resistant cultivars available, biorationals and biopesticides available etc. into account.

Cultural control

This entails several agronomic manipulations to make the cropping system less friendly to the pest and benefit the natural enemies, thereby, enhancing the natural control process. Maintaining and increasing the biological diversity of farm systems is the main strategy of cultural control. Decreased biodiversity tends to make the agro ecosystems more unstable and thus prone to recurrent pest outbreaks. Systems high in biodiversity tend to be more dynamically stable. This entails several agronomic manipulations to make the cropping system less friendly to the pest and benefit the natural enemies, thereby, enhancing the natural control process. Plant multiple varieties on multiple dates to diversify and control the risk of spread.

Host Plant Resistance

Insect pest resistant varieties are continually being bred by researches. But as natural systems are dynamic rather than static, breeding for resistance must be an ongoing process.

Prune and train properly

- Pruning and training fruit plants are cultural practices that can ensure good production.
- They also permit light, air, and spray materials to readily penetrate throughout the canopy.
- Pruning is an important technique for managing insect pests.

Fertilize appropriately

- Healthy plants are generally less susceptible to insect damage and produce more desirable products.
- Apply a balanced fertilizer according to recommendations. Avoid over-fertilization with nitrogen because that can cause rapid growth, which encourages certain insect pests.

Control weeds

- Managing weeds helps plants grow more vigorously and avoid insect and disease problems.
- Keep the grass around garden area and orchard plantings closely mowed, and keep the ground clean around the bases of trees. These practices will help control insects, diseases, and rodents.
- Weeds can be a real problem in small fruit plantings because most small fruits have shallow root systems, and weeds (especially grasses) can be very competitive. The soil around small fruit plants should be free of all vegetation.
- Hand pull or hoe weeds, or mulch. Mulching with black plastic, woven fabric, or organic materials (including grass clippings, leaves and bark chips) is one of the most effective ways to reduce weed problems in small fruit plantings.
- Around small fruit plantings, glyphosate is the only herbicide that should be used; however, you should always be careful not to get any of the herbicide on any green parts of the plants.

Mechanical and physical control

- Utilization of physical computes of environment like temperature, humidity, light to the detriment of the pest or killing the pest by some mechanical means come under mechanical and Physical control.
- Tillage, flaming, flooding, soil solarization by use of mulches etc.
- Deep summer ploughing exposes the stages of several insect pest to the harsh sun, thus killing them. This also make the hiding insect stages more prone to the attack of natural enemies and insect feeding birds like black drongo, sparrows, etc.
- Bag fruits - Fruit bagging is simple: when fruit is still small, place bags only around the fruits that will be consumed. The fruit remains bagged until three weeks before harvest so they can develop color and ripen properly.
- The bags are a protective barrier against attack by summer insect pests. When combined with resistant cultivars, bagging fruit can significantly reduce the number of pesticide applications.
- Use of light traps

Pheromone trapping

- Pheromone: An intraspecific message-carrying chemical are very effective means of managing the important insect pests of fruit crops, like fruitflies.

Biological Control

- It involves the use of living organisms (parasites, predator and pathogen) to maintain the pest populations below those causing economic damage.
- The first step in setting up a BIPM program is to assess the populations of beneficial organisms and their interactions within the local ecosystem.
- This helps in determining the potential role of natural enemies in the managed agricultural ecosystem.
- When pest are rationally kept at lower levels by the natural enemies present in the biodiverse system, it is referred to as natural biological control/natural control.

How to enhance the natural control system?

Creation of habitat to enhance the chance for survival and reproduction of beneficial organism i.e. habitat enhancement is termed farmscaping. We must be able to recognize the natural enemies and know what they do. Establishment of flowering annual / perennial plant, providing water, alternative prey, perching sites, overwintering sites, etc. for beneficial organisms. Augmentative or applied biological control involves supplementation of beneficial organism population by their periodic releases. Minimizing insecticide use, selective use of insecticides in selective ways, maintaining favourable habitats, providing alternate foods (pollen, nectar, etc.) augment the already available force of bioagents. While in case of inundative release immediate check on pest populations is targeted.

Several insecticides have been recommended for organic production. These include -

- Microbials
- Botanicals
- Soaps and Oils
- Abrasives
- Elemental compounds
- Pheromones
- Particle film barriers

Insect pathogens as microbial insecticides

- Bacteria - *Bacillus thuringiensis (various subspecies)* Many formulations of Bt are sold for control of caterpillars, eg. *Bacillus thuringiensis kurstaki and aizawai*. These are toxic only to Lepidopteran larvae (caterpillars), but it must be ingested to be effective. However, they are not effective against larvae that bore or tunnel into plants without much feeding on the surface (such as codling moth and oriental fruit moth). It is degraded by ultraviolet light and has short residual activity on treated foliage. Leaf rollers and other lepidopterous pests are the good targets in fruit crops. They are available commercially with varied names Dipel®, Agree®, XenTari®, and many others. Likewise, Viruses, Fungi - *Beauveria, Entomophthora, and Metarrhizium spp.,* Nematodes - *Steinernema & Heterorhabditis*, Spinosads - Derived from a soil actinomycete may be used for managing fruit pests in an effective way. Mycotrol® is a microbial insecticide containing *Beauveria bassiana.* All these are effective against a range of insects, including apple maggot (fair), oriental fruit moth, and codling moth.

Use of biorationals

- Although cultural controls dramatically reduce pest problems, the reality is that most plantings require pesticide applications to ensure a sufficient amount of clean, attractive fruit.
- Pesticides are merely another tool to manage problems. And like any tool, there are some basic things to know about pesticides before using them.
- Biorationals are generally considered to be derived from naturally occurring compounds on microorganisms' formulations.
- But they have narrow target range and are environmentally benign.
- For eg. *Bacillus thuringiensis*, nuclear polyhedrosis viruses, Granular viruses, various fungi like, *Beauveria bassiana*, *Metarrhizium anisopliae*, *Nomurae rileyi*.
- Biorational also include botanicals, silica aerogels, insect growth regulators, attractants, repellents and particle film barriers.
- The synthetic analogues of insect hormones that modify several growth responses in insects are referred to as insect growth regulators (IGRs).

Botanicals – are the insecticides derived from plants. For eg.

Pyrethrins - from pyrethrin daisies

- These are axonic poisons, i.e. act on the nervous system of the target insect.
- But are low in toxicity to mammals, thus ensuring safety to humans and other wildlife.
- They breakdown very rapidly so has no residual action and are environmentally safe.
- They are labeled for use on many fruits.

Neem

- It has been extracted from all parts of *Azadirachta indica* and *Melia* spp.
- This is low in toxicity to mammals.
- It has many uses medicinally.
- But has very short persistence, so are safe.
- They are labeled for use on many crops and fruit trees, especially against soft-bodied insects.

There are several other plants that can be used for the preparation of insecticide

- Castor (*Ricinus communis*)
- Darek (*Melia Azederach*)
- Lantana (*Lantana Camara*)
- Aak (*Calotropis procera*)
- Bhang (*Cannabis sativa*)
- Wild tobacco (*Nicotiana glauca*)
- Curry leaves (*Murraya koenigii*)
- Bana (*Vitex negundo*)
- Congress grass (*Parthenium hysterophorus*)
- Custard apple leaves / seeds
- *Adhatoda vesica*

Flow Chart showing preparation method of insecticide from various plant materials

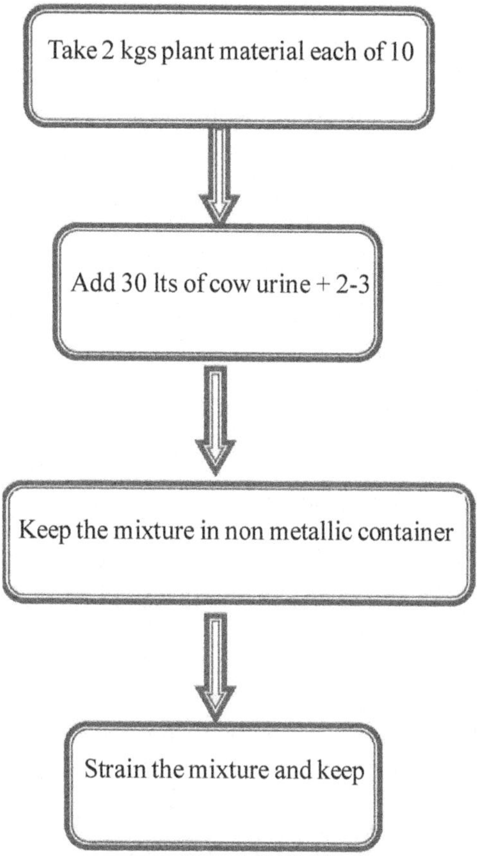

Method of preparation of NSKE (Neem Seed Kernel Extract)

11 kg neem seed kernel crushed well and mixed in 2 L water is kept for 24 h.

It is filtered and mixed with 20 L of water + soap water (100 ml)

It controls caterpillars, aphids, jassids, hoppers, etc.

Other techniques - Garlic (1kg) + Ginger (0.5 kg) + green chilli (0.5 kg) + 10 L water is left for one night. Use 0.5 to 1 L solution per tank along with 100 ml soap water.

Liquid pest repellents

Cow's urine + water (1:3)

Neem oil + water (1:3) may also be used.

Thin paste of cow dung, cow urine and clay can be applied on fruit tree trunk to manage pests like bark eating caterpillar, mealy bugs, etc.

Oils – Two types are frequently being used in fruit crop production - Dormant oils and Summer oils, against mites, aphids, other soft-bodied pests. But complete coverage is essential (upper and lower leaf surfaces). Oils kill by suffocating pests that are sprayed directly on to them.

Insecticidal soaps

- These are salts of fatty acids.
- They kill insects by disrupting membranes (including tracheal linings).
- But they work only against those insects that are wetted by the spray. It has no residual action.
- These are effective against aphids, whiteflies, mites, and other soft-bodied, not too- mobile pests.

Elemental and naturally occurring chemicals

- Sulfur
- Effective miticide (may cause plant injury)
- Copper
- Arsenic
- Lead arsenate and copper arsenite were among the earliest widely used insecticides

Abrasives - Clays, diatomaceous earth, silica aerogels disrupt the insect's cuticle and kill by dehydration. Kaolin clay comes with the trade name Surround®.

The active ingredient is kaolin clay, an edible mineral long used as an anti-caking agent in processed foods. It is applied as liquid, which evaporates leaving a protective powdery film on leaf surface. This film deters insects either by making them unsuitable for feeding and egg laying or less recognizable as a host. Sugar esters have been found more effective against mites and aphids in apple orchards, psylla in pear orchards, white flies, thrips and mites on vegetables and whiteflies on cotton, than conventional insecticides. These act as contact insecticides and degrade into environmentally benign sugars and fatty acids after application.

Fruit crops

Mango – It is attacked by several insect pests viz., mango hoppers, mealy bugs, stem borer, fruit fly, etc.

Management strategies to be adopted for effectively managing these insect pests are

- Remove and destroy dead and severely affected branches of the tree.
- Remove alternate host, silk cotton and other hosts.
- Grow tolerant mango varieties viz., Neelam, Humayudin.
- Swab Coal tar + Kerosene @ 1:2 (basal portion of the trunk - 3 feet height) after scraping the loose bark to prevent oviposition by adult beetles.
- Hook out the grub from the bore hole – apply Kerosene oil + Neem based insecticide 10 to 20 ml/ hole.
- Avoid close planting, as the incidence is very severe in overcrowded orchards.
- Orchards must be kept clean by ploughing and removal of weeds.
- Neem oil 5 ml/lit of water or neem seed kernel powder extract 5 per cent can be sprayed, first at the time of panicle emergence, second two weeks after first spray.
- Apply Buprofezin 25 EC @ 1.5 ml / L
- Collect fallen infested fruits and dispose them by dumping in a pit.
- Follow deep summer ploughing to expose the pupa.
- Monitor the activity of flies with methyl eugenol sex lure traps.
- Bait spray - combing any one of the insecticides and molasses or jaggery 10 g/l, two rounds at 2 weeks interval before ripening of fruits.
- Prepare bait with methyl eugenol 1% solution mixed with malathion 0.1%. Take 10 ml of this mixture per trap and keep them in 25 different places in one hectare
- Field release of natural enemies *Opius compensates* and *Spalangia philippines* against fruit fly.
- Remove weeds and grasses by ploughing during June-July.
- Band the trees with 20 cm wide alkalthene of polythene (400 gauge) in the middle of December (50 cm above the ground level and just below the junction of branching).

- Stem with jute thread and apply a little mud or fruit tree grease on the lower edge of the band.
- Release of Australian ladybird beetle, *Cryptolaemus montrouzieri* @ 10/tree

Guava

Mealy bugs, scales, fruit borers, fruit fly, bark eating caterpillar, white flies, etc. seriously inflict guava trees. These insect pests can be managed by adopting following management techniques.

- Installation of yellow sticky traps and removal of alternate host plants (*Ageratum conzoides*) for whiteflies.
- Same as in case of mango pests.

Citrus

Citrus psyllid, black aphid, fruit sucking moths, mealy bug, leaf miner, citrus scale, etc. are the major insect pests attacking citrus trees. The below mentioned techniques may be followed to effectively manage the insect pests.

- Collect and destroy the damaged plant parts.
- Encourage activity of parasitoids, *Encarsia* sp., *Eretomocerus serius* and *chlysoperla* sp. and predatore like Syrphids and Chrysopids.
- Field release of Australian lady bird beetle *Cryptoleamus montrouizeri* 10 per tree.
- Use sticky trap (5cm length) on fruit bearing shoots.
- Apply mixture of manure compost tea, molasses, citrus oil.
- Control ants and dust which can give the scale a competitive advantage.
- Field release of vedalia and Australian ladybugs.
- Destroy the weed host *Tinospora cardifolia* and *coccules pendules* of fruit sucking moth
- Bag the fruit with polythene bag (500 gauge)
- Apply smoke to prevent adult moth
- Trap crop – growing tomato crop in orchards to attract the adult moth
- Poison baiting with dilute suspension of fermented molasses and malathion 0.05% (50 EC at 1ml/lit)
- Use light trap or food lure to attract moths.

- Spray dormant oil in late winter before spring.
- Spray horticultural oil, if needed, year round.

Pomegranate

Anar butterfly, fruit borer, mealy bug, white flies, aphids are its major insect pests. The following strategies may be followed to manage these insect pests.

- Collect and destroy damaged fruits, plant parts, etc.
- Clean cultivation as weed plants serve as alternate hosts.
- Endemic areas - grow less susceptible varieties.
- Cover the fruit with polythene bags when the fruits are up to 5 mm.
- Use light trap @ 5/ ha to monitor the activity of adults.
- Flowering stage - spray NSKE 5% or neem formulations 2 ml/1.
- Release *Trichogramma chilonis* at one lakh/acre, first instar larva of *Chrysoperla carnea* @ 15 / flowering branch (four times) at 10 days interval from flower initiation during April., Coccinellid predator, *Cryptolaemus montrouzieri* and lace wing fly, *Mallada astur.*

Aonla

This fruit tree is inflicted by leaf roller, fruit borer, fruit sucking moth, stem borer, bark borer, mealy bugs and scales. These pests can be managed by following below mentioned techniques.

- Clean cultivation.
- Collection and destruction of infested plant parts along with leaf roller.
- Removal of weed plants
    - *Tinospora cordifolia*
    - *Cocculus pendules*

As they act as alternate host for fruit sucking moths. They also lay eggs on these weed species.

- Destruction of fallen and decayed fruits.
- Smoking.
- Collection of moths at evening by hand nets.
- Collection of semiloopers from the weeds and creepers.
- Use light trap - Destruction in kerosenised water below such light.
- Use of poison baits.

- Collect loose and damaged bark & destroy.
- Kill larvae by inserting iron spike or wire into hole.
- Field release of syrphids or green lacewing, *Chrysoperla carnea* can control aphid population rapidly.
- Spray neem oil at 3%.
- Early detection of mealy bugs - presence of ants – indicator.
- Cutting of infested twigs and leaves and burying them.
- Several species of ladybird beetles such as *Chilocorus* sp., *Cryptolaemus montrouzieri* are efficient predators.

Pests of Temperate Fruits

San Jose Scale, wooly aphid and apple codling moth are the major insect pests inflicting temperate region fruit trees.

Management strategies to be followed for managing them are

- Spraying the dormant trees in winter with 3% miscible oil (Dormant oil).
- In addition to spraying, the parasitoid, *Encarsia pernicoisi* (Tower) may also be released to check the over wintering population of San Jose scale on wild host plants growing around.
- Dormant oil: These are heavy and less refined oil suitable for application on fruit trees and shrubs during dormant season when they will be devoid of leaves. These are available in emulsifiable forms having 85 – 95% of actual oil.
- The aphid population can also be effectively checked by an exotic parasitoid, *Aphelinus mali Hald.*
- Strict domestic quarantine is to be followed by screening of consignments of fruits to prevent the spread of the apple codling moth from Laddak to other apple growing regions.
- Collect and destroy the infested fruits to prevent the carryover of the pest.
- Application of 0.2% Pyrethrum extract is also helpful in checking the pest infestation. The protective treatment may be applied about ten days before ripening of the fruits.

Conclusion

- The flexibility and environmental compatibility of IPM strategy make it useful in all types of ecosystems.
- There are several ecofriendly ways available to reduce the pesticide usage and produce optimization.
- There is a need to realize the potential of indigenous biocontrol agents and attention should be given to conserve them.
- Standardization of conservation methods for parasitoids and predators, use of pheromones and other biorational means of management should be intensified.

For the effective implementation of these techniques under BIPM, exact pest identification, regular monitoring, knowledge of economic threshold level and time and resources available for BIPM is essential.

References

http://agritech.tnau.ac.in/crop_protection/crop_prot_crop_insectpest%20_guava.html
https://www.google.co.in/?gfe_rd=cr&ei=U_72WM_ HAeTs8Aexu LeICQ#q=insect+pests+of+mango

9

Mass Trapping of Fruit Flies Using Methyl Eugenol Based Traps

Sandeep Singh and Kavita Bajaj

Fruit Flies and Economic Importance

Fruit flies belong to order Diptera (true flies) with one of the largest, most diversified and fascinating acalypterate family Tephritidae. These are commonly called as fruit flies due to their close association with fruits and vegetables and are also known as peacock flies because of their habit of strutting about and vibrating spotted and striped wings. Of the 4500 known species of fruit flies worldwide, nearly 200 are considered as pests but 70 species are regarded as agriculturally important throughout the world (Clarke *et al.*, 2005, Schutz *et al.*, 2012). David and Ramani (2011) reported 325 species in the Indian subcontinent of which 243 in 79 genera are from India alone. Important fruits flies damaging fruit crops in India include Oriental fruit fly or mango fruit fly, *Bactrocera dorsalis* (Hendel); peach fruit fly, *B. zonata* (Saunders); guava fruit fly, *B. correcta* (Bezzi); melon fly, *B. cucurbitae* (Coquillett) and *ber* fruit fly, *Carpomyia vesuviana* Costa, whereas important fruit flies in Punjab include *B. dorsalis* and *B. zonata*. Important host plants of *B. dorsalis* are mango, guava, peach, citrus, pear, *ber* and loquat but most preferred host of *B. dorsalis* is guava while that of *B. zonata* are guava, peach, mango, pear, *ber,* citrus and loquat but *C. vesuviana* infests only *ber* (Mann 1990, Sharma *et al.*, 2005, Singh 2008b, Singh 2012, Singh *et al.*, 2014a,b and Singh *et al.*, 2015). *Bactrocera dorsalis* is considered to be the most damaging and aggressive fruit flies in the world (Leblance and Putao 2000). They are strong fliers and can fly upto two kilometers in search of food (Butani 1979).

Fruit flies are of great economic importance as majority of them cause extensive damage to many fruits and vegetables and ruin more than 400 different fruit and vegetable crops including mango, guava, citrus, melon, papaya, peach, passion fruit, plum, apple and star fruit (White and Elson-Harris 1992). Besides fruit and vegetables crops, they are also destructive to many oilseed crops and ornamental plants. Mann (1980) advocated the seasonal history and occurrence of *B. dorsalis* on different fruit crops in Punjab whereas *B. dorsalis* and *B. zonata* have been found damaging Kinnow mandarin during August to October (Singh 2006, 2008 a,b,c, Singh *et al.*, 2013, Singh *et al.*, 2014a,b and Singh *et al.*, 2015). Sharma *et al.*, (2008) reported that fruit fly infestation was more on central guava trees than on the peripheral trees. Fruit fly damage also increased with the increase in size and maturity of guava fruits.

Losses due to fruit flies occur due to decrease in production by direct damage on fruits and vegetables, increase in management cost and restricting free trade and movement of fruits from the countries of fruit fly prevalence to fruit fly free zones. Apart from causing direct yield losses, fruit flies also cause major economic impact especially through quarantine and regulatory programmes, costly survey and field control strategies, eradication programmes, disinfestations treatments and prevention of development of desirable food crops (Christenson and Foote 1960, Bateman 1991, White and Elson-Harris 1992, Abdullah *et al.*, 2007, Verghese *et al.*, 2012). Due to their infestations, India has been included in the list of those countries from where fruit import to some of the developed countries is banned (Stonehouse 2001).

The fruit flies are also very difficult to manage due to the fact that they are polyphagous, multivoltine, adults have high mobility and fecundity and all the development stages are unexposed (Prokopy 1977, Vargas *et al.*, 1984, Sharma *et al.*, 2011) and adult flies may live for 3 months (Mann 1990). Two important parameters including the life-history strategies of fruit flies are favourable environment for reproduction, survival and host availability in time and space (Bateman 1972, Fletcher 1989). Present management strategies mostly focus on chemical insecticides. Due to cryptic nature of the maggots and eggs of fruit flies, they mostly remain unaffected by such insecticides and only adults are exposed to control measures. So, most of insecticidal treatments are ineffective. Furthermore, application of insecticides disrupts the ecosystem and causes numerous hazards, which in present scenario warrants the need of integrated approach for fruit fly management. Sanitation combined with the use of lures and traps as well as baits proved to be the best alternatives for management of fruit flies. Traps having chemical cues and signals which influence the behaviour, physiology, and ecology of fruit flies in a remarkably large number of ways are used for variety of reasons like suppression, surveillance and ecological study. This strategy is itself diverse and may involve the elimination, modification, disruption and imitation (Shelly *et al.*, 2014). The traps have high specificity, low cost and are environmentally quite safe (Sureshbabu and Viraktamath 2003, Ravikumar 2005, Singh and Sharma 2013). Among the various alternate strategies available for the management of fruit flies, the use of methyl eugenol based traps stands as the most outstanding alternative.

Methyl Eugenol- Parapheromone

Tephritid fruit flies show strong behavioural responses. Methyl eugenol is widely recognized as the most powerful male lure currently in use for detection, control and eradication of any tephritid species (Drew 1974, Hardy 1979, Drew and Hooper 1981, Kapoor *et al.,* 1987, Drew and Hancock 1994, Verghese *et al.,* 2006, Vargas *et al.,* 2009b, Singh and Sharma 2011, Singh *et al.,* 2011, Kumar and Sharma 2012, Sharma 2012, Verghese *et al.,* 2012, Singh and Sharma 2013, Singh *et al.,* 2014a,b and Singh *et al.,* 2015. It is more correctly called methoxy eugenol (4-allyl-1,2-dimethoxybenzene) (Fig 1). Trapping is a useful tool that offers a lot of possibilities to the study and control of fruit flies. The species present in a specified area can easily be catalogued by determining their geographic situation, seasonal abundance, host status and monitoring of already established fruit fly populations (Allwood 1997). The information collected in traps such as number of flies and species, is a valuable source for not only the planning of control programmes but also for quarantine detection.

Methyl eugenol is parapheromone, which is defined as chemical compound of arthropogenic origin, not known to exist in nature but structurally related to some natural pheromone components that in some way affects physiologically or behaviourally the insect pheromone communication system (Renou and Guerrero 2000). It has both olfactory as well as phagostimulatory action and it is known to attract fruit flies from a distance of 800 m (Roomi *et al.,* 1993, Bhagat *et al.,* 2013, Haq *et al.,* 2014). Methyl eugenol was used already in early 1900 (Howlett 1912) and its effectiveness in attracting *B. dorsalis* has been well documented. Methyl eugenol attracts male flies (e.g. from upto 500 m away) but not the female flies (Kumar 2011). When population of fruit flies is more, even female flies are also attracted towards methyl eugenol (Verghese 1998). It specially attracts the males of *B. dorsalis, B. correcta* (Bezzi) and *B. zonata* (Verghese *et al.,* 2006). There is evidence that methyl eugenol is involved in the mating effectiveness where males that feed on this lure increase mating success (Shelly and Dewire 1994).

Methyl eugenol, when used together with an insecticide (malathion, fipronil or naled) impregnated into a suitable substrate, forms the basis of male annihilation technique (MAT) and results in the reduction of male population of fruit flies to such a level that eradication and suppression is achieved (Vargas *et al.,* 2010a). This technique has been successfully used for the eradication and control of several *Bactrocera* species (Cunningham 1989, Singh 2012, Singh and Sharma 2013, Singh *et al.,* 2014a, Singh *et al.,* 2014b, Singh *et al.,* 2015). Metcalf (1990) reported that atleast 58 species of *Bactrocera* are attracted to methyl eugenol. Methyl eugenol attracts males of many *Bactrocera* species upto 500 m (Ferrar 2010), but not members of the subgenus *Bactrocera* (*Zeugodacus*) which includes *B. cucurbitae,* and also *B. caudata* Fabricius and *B. tau* (Walker).

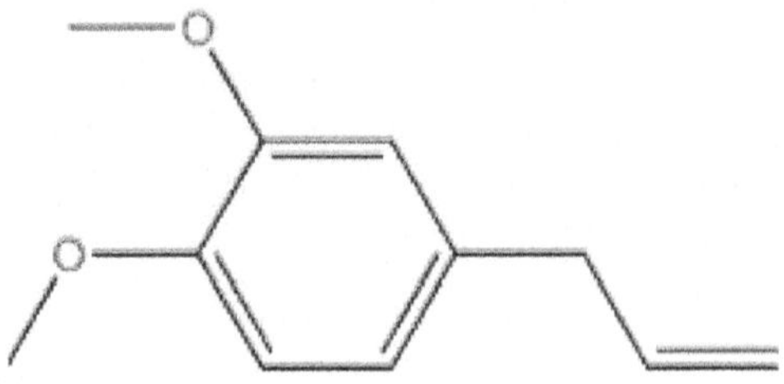

Fig. 1: Methyl Eugenol

Management Strategies through Mass Trapping

Evaluation of Methyl Eugenol based Traps

Tephritid fruit flies (*Bactrocera* spp.) being serious pests of orchard fruits throughout the world, may be controlled by MAT (Cunningham 1989) and has been successfully used in South Asia on guava (Marwat *et al.,* 1992). Boller (1983) observed that continuous use of attractant leads to annihilation of male flies, thus their chances of mating with the females become less and the same reduces the production of further progeny. Metcalf (1990) reported that atleast 176 species of *Bactrocera* are attracted to cue-lure and 58 to methyl eugenol. Verma and Nath (2006) has reviewed the research work conducted on different aspects of the management of the fruit flies through trapping like bait sprays, traps and lures.

It was reported that one per cent methyl eugenol alongwith 0.5 per cent malathion or 0.1 per cent carbaryl was most effective against *B. dorsalis* (Balasubramaniam *et al.,* 1972, Lakshmanan *et al.,* 1973). They also advocated monthly replenishment of methyl eugenol. However, Cunningham *et al.,* (1975) used 83 per cent methyl eugenol alongwith 10 per cent naled and 7 per cent thixein for management of *B. dorsalis*. Methyl eugenol (0.025, 0.05, 0.075 and 0.1 ml) impregnated on 2 cm^2 cotton wad had no significant difference in their efficacy when replenished at weekly interval (Belavadi 1979). Under field conditions, it was observed that single application of methyl eugenol at 0.075 ml was most effective upto 17 days for capturing *B. dorsalis* if the population was low (0-22 fruit flies/trap) and upto 32 days when the population was high (0-81 fruit flies/trap).

Bose *et al* (1979) used an attractant and insecticide solution containing 0.1 per cent methyl eugenol, 0.1 per cent dichlorvos (DDVP) and 0.5 per cent acetic acid in water against *B. dorsalis* on guava. Average rate of infestation of the pest was 12 per cent. Liu (1991) showed that 10 per cent (mixture of 10% methyl eugenol and 90% cue-lure) and 20 per cent MC (mixture of 20% methyl eugenol and 80% cue-lure) attracted more number of fruit flies than methyl eugenol alone. It was also observed that 10 per cent MC was effective lure for *B. cucurbitae* (Liu and Lin 1992). Madhura (2001) reported that 100:0 and 90:10 ratios of methyl eugenol and cue-lure attracted the maximum number of fruit flies compared to other proportions. Singh (1993) reported a significant reduction in the *B. dorsalis* population by using 0.1 per cent methyl eugenol baited traps in guava orchard. According to Makhmoor and Singh (1998), 1 per cent concentration of methyl eugenol was significantly superior to all other treatments for the control of *B. dorsalis* in guava orchard. Agarwal and Kumar (1999a) found mango pulp mixed with methyl eugenol as the most effective formulation against *B. zonata.* The use of coloured plastic open pan traps with methyl eugenol (0.1%) and 1.25 g carbofuran 3G in mango orchards revealed that white and yellow traps, placed on ground in the periphery of orchard were effective (Sarada *et al.,* 2001). It was recommended to use 0.1 per cent solution of methyl eugenol for trapping fruit flies in mango and guava (Anonymous 2004).

Eradication/suppression campaigns were made by using combination of methyl eugenol and insecticides against *B. dorsalis* (Steiner *et al.,* 1965, Bindra and Mann 1978, 1979, Ushio *et al.,* 1982, Iwahashi 1984, Mann 1986, Verghese *et al.,* 2004, Stonehouse *et al.,* 2007, Vargas *et al.,* 2008a,b, 2009a,b, 2010a,b). *Bactrocera dorsalis* infesting guava and other fruits in Japan was successfully eradicated from Okinawa Island by MAT (Koyama *et al.,* 1984). Until the number of males caught in monitoring traps was reduced to about 0.001 per cent of that before control, no detectable reduction of infestation level of host fruits was found. In Taiwan, fine fibreboard was found to be the most attractive dispenser of methyl eugenol 6 weeks after application to control males of *B. dorsalis* (Chu *et al.,* 1985). The fruit fly was effectively controlled on 59.1-79.4 per cent area of Lamby Island in Taiwan as no fruit damage was found in a male annihilation trial with fibreboard soaked in methyl eugenol (95%) and DDVP (92%) (Chiu and Chu 1988). In Pakistan, 80.48 per cent reduction in trap catch of *B. dorsalis* males were observed one week after the Eugecide S (methyl eugenol) baited traps when placed in guava orchards, 5 feet above the ground at a density of 1 trap/acre (Marwat *et al.,* 1992).

Patel *et al.*, (2005b) observed that MAT traps containing plywood blocks caught maximum number of fruit flies (710 flies) compared with traps made from wood of twenty one different trees or sources, whereas Patel *et al.* (2005a) found ethanol as best solvent for use in MAT alongwith methyl eugenol and DDVP in the ratio of 6:4:1 (ethanol: methyl eugenol: DDVP). Stonehouse *et al.* (2002b) reported that plywood blocks attracted and killed more flies than those of mulberry and poplar wood blocks. Moreover, square and oblong blocks were more effective than round and hexagonal blocks. MAT using methyl eugenol traps @ 4 traps/acre in mango and guava has been found to be very effective in controlling fruit flies in different parts of India (Stonehouse *et al.*, 2005). Verghese *et al.* (2006) reported that MAT @ 4 traps/acre (plywood impregnated with methyl eugenol and dichlorvos) alongwith sanitation (removal and destruction of fallen fruits every week) appreciably brought down *B. dorsalis* infestation in mango, whereas Viraktamath and Ravikumar (2006) observed maximum number with 16 traps per acre followed by 8 traps. The per cent infested fruits and level of maggot incidence declined to zero level in these treatments as against 42.33 to 64.29 per cent infested fruits and 1.43 to 3.75 maggots/fruit in control. Ravikumar and Viraktamath (2006) conducted an experiment on number of holes in an 1000 ml capacity pet bottle trap containing methyl eugenol in guava and mango orchards. Bottle traps with 4 holes of 20 mm diameter were found significantly superior in attracting higher number of adults of *B. dorsalis*, *B. correcta* and *B. zonata* than those with 1, 2, 3, 5 or 6 holes/ trap.

Data collected by Singh *et al.*, (2007) from methyl eugenol based traps in Uttar Pradesh, revealed that 5 species of fruit flies namely *B. zonata*, *B. affinis* Hardy, *B. dorsalis*, *B. correcta* and *B. diversa* (Coquillett) were attracted. Maximum number of fruit flies (571.0/trap) was trapped in 21st standard week during 2005. Shankar *et al.* (2010) conducted field studies in Andhra Pradesh to determine the effect of the number and size of holes on traps on the capturing efficiency of *Bactrocera* spp. (*B. dorsalis*, *B. correcta* and *B. zonata*) in mango. Results showed that higher fruit fly numbers were recorded in the treatment bottle with 4 holes with 20 mm size (9.37 flies/trap/week) followed by the trap with one hole (9.33 flies/trap/week). The influence of hole size revealed that fruit fly adults were attracted to the traps with 8 mm diameter holes (9.33 flies/trap/week), which was attributed to the quick dispersal of methyl eugenol and the influence of weather factors.

The surveillance of *B. invadens* in parts of South Africa through trapping with methyl eugenol helped in early detection and eradication of this fly a possibility (Manrakhan *et al.*, 2009). The re-invasion of Okinawa, Japan by the *B. dorsalis* complex after its eradication was comprehensively documented for the first time by Ohno *et al.* (2009). From 1987 to 2008, more than 300 adult flies were captured by monitoring traps baited with methyl eugenol, showing frequent re-invasion. During this period, re-colonization (detection of infested fruits) occurred six times and in all cases the flies were successfully re-eradicated by countermeasures (mainly MAT). Singh *et al.* (2009) carried out an investigation to evaluate the performance of eco-trap, methyl eugenol bait and carbaryl against fruit fly in mango in Orrisa. Maximum number of fruit flies were trapped in treatment (25 trees/eco-trap) followed by 20 trees per eco-trap and 30 trees per eco-trap. In methyl eugenol treatment, the number of fruit flies trapped was 34.74, 42.05 and 28.98 at 5, 10 and 15 days after installation of trap, respectively. Effectiveness of the eco traps was maintained upto 45 days. Methyl eugenol and cue-lure traps to detect tephritid flies on the U.S. mainland were tested by Vargas *et al.* (2009a) with and without insecticides under Hawaiian weather conditions against small populations of *B. dorsalis* and *B. cucurbitae*, respectively. In comparative tests, standard Jackson traps with naled and the Hawaii fruit fly area-wide pest management (AWPM) trap with DDVP insecticidal strips outperformed traps without an insecticide. It was concluded that Farma Tech methyl eugenol and cue-lure wafers with DDVP would be more convenient and safer to handle than current liquid insecticide formulations (e.g. naled) used for detection programmes in Florida.

Vargas *et al.* (2009b) conducted studies in Hawaii, USA to quantify attraction and feeding responses resulting in mortality of male *B. dorsalis* to a novel MAT formulation consisting of specialized pheromone and lure application technology (SPLAT) in combination with methyl eugenol (ME) and spinosad (=SPLAT-MAT-ME with spinosad) in comparison with Min-U-Gel-ME with naled (Dibrom). The results indicated that SPLAT-MAT-ME with spinosad offers potential for control of males in an area-wide IPM system without the need for conventional organophosphates. Singh *et al.* (2009) carried out an investigation to evaluate the performance of eco-trap, methyl eugenol bait and carbaryl against *B. dorsalis* in mango. Maximum number of fruit flies were trapped in treatment (25 trees/eco-trap) followed by 20 trees per eco-trap and 30 trees per eco-trap. In methyl eugenol treatment, the number of fruit flies trapped was 34.74, 42.05 and 28.98 at 5, 10 and 15 days after installation of trap, respectively. Effectiveness of the eco-traps was maintained upto 45 days.

Casagrande (2010) discussed the principles and design of the Magnet MED system (attract and kill) for the control of the *Ceratitis capitata* (Wiedemann)and describes the efficacy of this method against this pest on citrus and peach in Italy and Spain. It was found that deltamethrin, and ammonium acetate and trimethylamine attractants were vital components of the control system. A study aimed to explore the longer tails of the dispersal of many tephritids indicated that many flies were recovered at unprecedented long distances (between 2-

11.39 km) from the release point (Froerer *et al.*, 2010). These long-distance recaptures aid in understanding the long tails of spatial distribution of fly movement that has been suspected of this species. Shelly *et al.* (2010) released males of *B. dorsalis* and *B. cucurbitae* within the detection trapping grid operating in southern California with the objective to measure the distance-dependent capture probability of marked males. Methyl eugenol was the more powerful attractant, and based on the mark-recapture data, it was estimated that *B. dorsalis* populations with as few as 50 males would always (>99.9%) be detected using the current trap density of five methyl eugenol-baited traps per 2.6 km^2 (1 $mile^2$). By contrast, they estimated that certain detection of *B. cucurbitae* populations would not occur until these contained 350 males.

Palam Trap, a lure based mineral water bottle trap was found effective in monitoring and management of 10 species of fruit flies including *B. dorsalis* and *B. zonata* in fruits and vegetables in Himachal Pradesh (Mehta *et al.*, 2010). Han *et al.* (2011) monitored the population dynamics of *B. dorsalis* using methyl eugenol-baited traps in China. Adults were captured from early July to the end of December in a citrus orchard and peaked in October and early November. Adult population peak coincided with the ripeness period of sweet oranges in October. Field surveys indicated that pear was the first host plant infested by *B. dorsalis* and recorded the following host shift pattern, i.e. pear (*Pyrus communis*), jujube (*Zizyphus jujuba*), persimmon (*Diospyros kaki*), and sweet orange (*C. unshiu*). The availability of preferred host fruits and the low winter temperature were key factors influencing population fluctuations.

Singh and Sharma (2011) reported usefulness of methyl eugenol based mineral water bottle traps in mass trapping of fruit flies, *B. zonata* and *B. dorsalis* on Kinnow in Punjab. Singh *et al.* (2011) recorded adult male fruit flies in methyl eugenol based traps at weekly interval in pear in Punjab. Trap catch ranged from 74.9 to 326.4 flies/trap/week during different years. Trap catch during the year 2006 ranged from 76.3 flies in first week of June to 326.3 flies in the fourth week of July. In 2007, trap catch varied from 78.4 to 300.8 flies while during 2008, the catch ranged was 70.2 to 352.3 flies. Results showed that the using traps with MAT can reduce the insecticide usage in pear orchards and help in better management of fruit flies, which otherwise was very difficult with insecticides.

Singh and Sharma (2012) opined that 16 traps/acre based on water bottle using methyl eugenol through MAT had significantly more population of male fruit flies, *B. dorsalis* and *B. zonata* as compared to 4, 8 and 12 traps/acre in peach. Similar results were found in rainy season guava (Singh and Sharma 2013). Reji Rani *et al.* (2012) reported that per cent infestation of mango can be significantly controlled by using methyl eugenol based traps @ 1 trap/0.1 ha. Sharma (2012) tested the efficacy of an improved form of mass trapping method using spinosad as an insecticide in methyl eugenol trap for the control of *Bactrocera* complex on mango, guava, sapota and peach at New Delhi. The catches of fruit fly increase having a maximum capture of 179 flies/trap during 2007 and 14.8 flies/trap in 2010, thus indicating 90 per cent reduction in mean capture/trap in 3 years and 6 months. Kumar and Sharma (2012) have advocated the use of methyl eugenol based traps for management of fruit flies in citrus. Verghese *et al.* (2012) also highlighted the importance of MAT in area-wide control of *B. dorsalis* infesting mango in South India. Singh and Sharma (2013) reported that methyl eugenol based traps have eco-friendly approach with great advantages like no labour cost, cheap as compared to chemical insecticides, insecticide residue free fruits and no ill-effect on natural enemies, human health and environment.

Singh *et al.* (2013) reported that the plyboard blocks impregnated with a mixture of ethanol, methyl eugenol and malathion (6:4:1) were found effective and persistent than suspension traps in trapping fruit flies in mango orchards in Himachal Pradesh. Nagaraj *et al.* (2014a) reported that 0.4 ml methyl eugenol was superior in attracting large number of *B. dorsalis* and *B. correcta,* while the highest number of *B. zonata* was recorded in the traps charged with 0.6 ml followed by 0.2 ml, 0.8 ml and 1 ml methyl eugenol, in Bengaluru.

Singh *et al.* (2014a) conducted an experiment on management of fruit flies, *B. dorsalis* and *B. zonata* in pear by fixing PAU fruit fly traps @ 16 traps/acre using methyl eugenol at different locations in Punjab and reported that fruit flies can be effectively managed in pear orchards. Similar results were also obtained in guava (Singh *et al.* 2014b) and in Peach (Singh *et al.*, 2015). The composition of different lures with commercial insecticides is presented in Table 1.

Role of colour in traps in capturing fruit flies

Greany *et al.* (1978) conducted an experiment in guava orchards in Miami, USA and found that fluorescent orange sticky traps reflecting maximally at 590 nm captured significantly more Caribbean fruit fly, *Anastrepha suspense* (Loew) than non-fluorescent orange traps. Fluorescent traps with peaks at 510 and 610 nm also tended to capture more flies than plain orange traps. Fly capture rates were directly proportional to total light reflected in the 580-590 nm regions and to the intensity of light of this hue. According to Nakagawa *et al.* (1978), 7.5 cm spheres of black and yellow coloured traps captured more females of *C. capitata* while black, red, orange and yellow traps captured more male in fruiting coffees at Hawaii. Cytrynowicz *et al.* (1982), in his experiments conducted in Brazil, reported that South American fruit flies, *Anastrepha*

Table 1: Composition of trap lure and fruit fly species trapped

Composition of Trap Lure	Fruit fly species	Crop	Reference
ME (1%) and Malathion (0.5%) or Carbaryl (0.1%)	*B. dorsalis*	Mango	Balasubramaniam *et al* 1972, Lakshmanan *et al* 1973
ME (83%), Naled (10%) and Thixein (7%)	*B. dorsalis*	Mango	Cunningham *et al* 1975
ME (0.4 ml) and dichlorovos (1 ml)	*B. dorsalis* and *B. zonata*	Orange	Ravikumar and Viraktamath 2006
ME (0.4 ml) and dichlorovos (1 ml)	*B. dorsalis* and *B. zonata*	Guava	Viraktamath and Ravikumar 2006
ME: Malathion (8:1)	*B. dorsalis, B. correcta* and *B. zonata*	Mango	Shanker *et al* 2010
Ethanol: ME: Malathion (6:4:1)	*B. dorsalis* and *B. zonata*	Mango and Guava	Stonehouse *et al* 2002b, Stonehouse *et al* 2005
Ethanol: ME: Malathion (6:4:1)	*B. dorsalis* and *B. zonata*	Kinnow	Singh and Sharma 2011
Ethanol: ME: Malathion (6:4:1)	*B. dorsalis* and *B. zonata*	Peach	Singh and Sharma 2012
Ethanol: ME: Malathion (6:4:1)	*B. dorsalis* and *B. zonata*	Mango	Reji Rani *et al* 2012
Ethanol: ME: Malathion (6:4:1)	*B. dorsalis* and *B. zonata*	Guava	Singh and Sharma 2013
Ethanol: ME: Malathion (6:4:1)	*B. dorsalis* and *B. zonata*	Mango	Singh *et al* 2013
Ethanol: ME: Malathion (6:4:1)	*B. dorsalis* and *B. zonata*	Pear	Singh *et al* 2014a
Ethanol: ME: Malathion (6:4:1)	*B. dorsalis* and *B. zonata*	Guava	Singh *et al* 2014b
Ethanol: ME: Malathion (6:4:1)	*B. dorsalis* and *B. zonata*	Peach	Singh *et al* 2015
ME (0.4 ml) and monocrotophos (1 ml)	*B. dorsalis* and *B. correcta*	Mango	Nagaraj *et al* 2014a
ME (0.6 ml) and monocrotophos (1 ml)	*B. zonata*	Mango	Nagaraj *et al* 2014a

fraterculus (Wiedemann) and *C. capitata* are more attractive toward yellow rectangles than orange, green or red ones and led to conclusion that colour attraction is more powerful tool in the trapping device. Green, yellow and orange were the most attractive colours for the Mexican fruit fly, *A. ludens* (Loew) in grapefruit (Robacker *et al.*, 1990). Vargas *et al.* (1991) and Stark and Vargas (1992) reported that *B. dorsalis* showed greater preference toward orange, yellow and white (transparent) colour in guava.

Epsky *et al.* (1995) found highest capture of *C. capitata* in green 3 hole traps with dull exteriors and with 12-15 cm width visual cue than on shiny orange traps on sweet orange (*C. sinensis*). In Palin, Guatemala, Heath *et al.*, 1995 conducted tests on clear traps versus traps with a painted coloured strips (~7.5cm high) around the periphery of the middle which were baited with a 2-component blend of ammonium acetate and putrescine to provide a visual cue. More females and males of *C. capitata* were captured in green and yellow traps, respectively on sweet orange whereas in case of *A. ludens*, neither male nor female flies differentiated among orange, green and yellow traps but the percentage trapped in any coloured trap was higher than in colourless traps.

Heath *et al.* (1996) observed that width of visual cue affected percentage capture of *C. capitata* females and traps with 12-15 cm width green visual cue captured more female fruit flies on sweet orange. Greater captures of *C. capitata* were achieved with orange and yellow bucket traps and orange modified bucket traps (Uchida *et al.*, 1996). Jalaluddin *et al.* (1998) found that *B. correcta* was more readily attracted to yellow and orange target than to red, green, white, violet or blue on guava. Liburd *et al.* (1998) observed that baited green, red, yellow or blue spheres were more attractive to blue berry maggot, *Rhagoletis mandex* (Walsh) than baited yellow board traps in V-orientation on blueberries in New Jersey. Cornelius *et al.* (1999) reported that yellow coloured spheres and rectangular blocks are more attractive to fruit flies as compared to red one on guava. Alyokhin *et al.* (2000) opined that effective lure-and-kill trap is a potentially important instrument in monitoring and controlling *B. dorsalis* on guava. Mayer *et al.* (2000) conducted an experiment on cherries and obtained more trap catch of western cherry fruit fly, *R. indifferens* Curran on 10 cm red spheres followed by 8 and 12 cm compared to vertical and V-oriented yellow rectangles. Seven different colours were studied for the capture of olive fruit fly, *B. oleae* (Rossi) by Katsoyanns and Kouloussis (2001) in Chios Island, Greece. They reported that yellow and orange 70 mm diameter plastic spheres coated with adhesive, trapped the greatest number of male *B. oleae* while red and black sphere trapped more females whereas white and blue spheres were the least effective for both sexes this was due to the fact that males were mostly captured by coloured spheres reflecting maximally between 580 and

600 nm (yellow to orange) with peak response at 590 nm and females were mostly captured by colours reflected maximally between 610 and 650 nm (orange to red) with peak response at 650 nm. Sarada *et al.* (2001) used coloured plastic open pan traps with methyl eugenol (0.1%) and 1.25 g carbofuran 3G in mango orchards and found that white and yellow traps placed on ground in the periphery of orchard were effective. Madhura (2001) observed that *B. dorsalis* showed greater preference toward orange, yellow and white (transparent) colour.

Jhala *et al.* (2005) reported that on little gourd, traps containing cue-lure and methyl eugenol attract most of fruit flies when coloured yellow, green, clear and white than blue, orange and red, and fewest when grey and blue. Jiji *et al.* (2005) found that *Bactrocera* spp. was significantly attracted to traps of increasing scale of redness with a peak at yellow/orange rather than red itself on all gourds and melon. Rajitha and Viraktamath (2005a) observed that red sphere (10.13/trap/day) captured more number of *B. correcta* than green cylinders whereas in case of *B. zonata,* transparent bottle trap (0.99/trap/day) captured more numbers as compared to red bottles (0.88/trap/day) in guava orchards in Karnataka. Rajitha and Viraktamath (2005b) opined that *B. dorsalis* showed greater response to green medium sized and all size orange spheres (0.45 to 0.49 fruit flies/trap/day). Big green and orange spheres were attractive to *B. correcta* (9.39-10.02 fruit flies) while *B. dorsalis* was attracted to big red, green and orange spheres and medium and big transparent bottle and medium red bottle (0.58-0.61 fruit flies) in mango orchards in Karnataka. Further, Rajitha and Viraktamath (2006) reported that in guava orchard, overall mean catches of *B. dorsalis* were significantly high in orange (31.2 fruit flies) followed by transparent traps (23.8 fruit flies) and mean trap catches in green and red traps was on par with each other (13.60 and 12.80 fruit flies, respectively) while *B. correcta* were high in orange (154.27 fruit flies) followed by red traps (118.93 fruit flies). Transparent and green traps were on par with each other (60.47 and 59.6 fruit flies, respectively).

Ravikumar and Viraktamath (2007) reported that yellow and transparent traps attracted significantly high number of *B. correcta* in guava (70.45 fruit flies/trap/week) and mango (5.13 fruit flies/trap/week), respectively whereas green and orange coloured traps in guava attracted 3.79 and 3.75 fruit flies/trap/week, respectively. Black coloured traps in mango (3.88 fruit flies/trap/week) were attractive to *B. dorsalis. Bactrocera zonata* were attracted to red coloured traps (3.75 fruit flies/trap/week) in mango ecosystem. When total fruit flies irrespective of species were considered, yellow colour traps were attractive in guava (71.91 fruit flies/trap/week) while black colour traps were effective in mango (8.68 fruit flies/trap/week).

Verghese and Mumford (2010) showed that trap colour influenced the capture of *B. zonata and B. correcta.* For *B. dorsalis* and *B. correcta,* the yellow traps clearly had a higher proportion of attraction followed by white and green whereas for *B. zonata*, results were equally good for yellow and black in mango orchards in Gujarat. Ros and Seris (2010) opined that transparent Easy blister trap and yellow Easy trap has equal efficacy for capturing *B. oleae* on olive in Spain. Yee (2011) reported that *R. indifferens* were caught on the Alpha Scents than Pherocon traps because of their different yellow colour and/or lower fluorescence and not the hot melt pressure sensitive adhesive (HMPSA) on cherries in Washington, USA. Yee (2013) conducted an experiment to capture *R. indifferens* by red and yellow sticky sphere traps on cherries and reported that yellow sphere capture more flies as compared to red traps.

Daniel *et al.* (2014) conducted an experiment on cherries in USA to attract European cherry fruit fly, *R. cerasi* (Linnaeus) on five different coloured yellow panels i.e. Panel I which was made up of polypropylene (0.8 mm) and fluorescent, Panel K which was made up of polypropylene (0.5 mm) and fluorescent, Panel F which was made up of polyethylene (1 mm) and not fluorescent, Panel G which was made up of polyethylene (1 mm) and fluorescent and Panel H which was made up of polyethylene (1 mm) and fluorescent and were compared to a standard Rebell® Amarillo trap. Results showed that Trap F, with a strong increase in reflectance at 500-550 nm and a secondary peak in UV-region at 300-400 nm captured significantly more flies than standard Rebell® Amarillo trap.

Nagaraj *et al.* (2014b) evaluated the efficacy of different coloured traps in capturing fruit flies in mango orchards at Bengaluru. When the total fruit flies irrespective of species were considered, yellow traps attracted more number of fruit flies with the mean trap catches of 18.60 fruit flies/trap/week followed by transparent and green colour traps (8.40 and 7.00 fruit flies/trap/week) which was on par with orange colour traps (4.80 fruit flies/trap/week). The lowest number of fruit flies was captured in orange colour traps. Jamwal *et al.* (2015) conducted an experiment in order to optimize the use of trap colour in lure charged traps. The studies were made on the extent of attraction between *B. dorasalis* and *B. zonata* in mango orchards of Saharanpur, Uttar Pradesh. Their results revealed that green coloured traps captured maximum number of fruit flies followed by orange traps and minimum flies were trapped in red followed by black coloured traps. Yee (2015) conducted an experiment on cherries to compare the capturing efficacy of yellow sticky stripes and yellow sticky rectangular boards and reported that yellow sticky stripes could be more

Table 2: Fruit fly species in response to trap colour

Crops	Composition of Trap Lure	Fruit fly species	Trap Colour	Reference
Mango	Ethanol: ME: DDVP (6:4:1)	*B. dorsalis*	Yellow, Orange and Transparent	Madhura 2001
Gourds and Melons	Instead of lure, 'Stickum' trapping gum	*Bactrocera* spp.	Yellow and Orange	Jiji *et al* 2005
Guava	Ethanol: ME: DDVP (6:4:1)	*B. correcta*	Red	Rajitha and Viraktamath 2005a
Guava	Ethanol: ME: DDVP (6:4:1)	*B. zonata*	Transparent	Rajitha and Viraktamath 2005a
Mango	Ethanol: ME: DDVP (6:4:1)	*B. dorsalis* and *B. correcta*	Green andOrange	Rajitha and Viraktamath 2005b
Mango	Ethanol: ME: DDVP (6:4:1)	*B. zonata*	Red,	Rajitha and Viraktamath 2005b
Guava	Ethanol: ME: DDVP (6:4:1)	*B. dorsalis* and *B. correcta*	Orange	Rajitha and Viraktamath 2006
Guava and Mango	Ethanol: ME: DDVP (6:4:1)	*B. dorsalis, B. correcta* and *B. zonata*	Yellow and Transparent	Ravikumar and Viraktamath 2007
Mango	Ethanol: ME: DDVP (6:4:1)	*B. dorsalis* and *B. correcta*	Yellow	Verghese and Mumford 2010
Mango	Ethanol: ME: DDVP (6:4:1)	*B. zonata*	Green	Verghese and Mumford 2010
Mango	ME (0.4 ml) and monocrotophos (1 ml)	*B. dorsalis, B. correcta* and *B. zonata*	Yellow	Nagaraj *et al* 2014b
Mango	Ethanol: ME: Malathion (6:4:1)	*B. dorsalis* and *B. zonata*	Green	Jamwal *et al* 2015

useful for capturing *R. indifferens* than yellow sticky rectangular boards in Washington, USA. The composition of different lures with commercial insecticides are summarised in Table 2.

Role of shape of traps in capturing fruit flies

Boller (1969) opined that spheres might be more attractive as they could be visible by the fruit flies from all direction. The biological basis for the acceptance of spheres over the other shapes may be due to similar shape of the oviposition hosts. Most of the earlier work was concentrated on comparing the response of *A. suspense* (Greany and Szentesi 1979) to different shapes of traps. Greany *et al.* (1977) and Sivinski (1990) reported that bigger orange rectangle and spheres were preferred by *A. suspense.* Prokopy (1973) demonstrated that more apple maggots, *R. pomonella* (Walsh) were captured on fluorescent yellow rectangles and enamel red spheres than other shapes of different colours. He hypothesized that flat surface of the rectangle represented leaf type stimulus, whereas spheres constitute a fruit type stimulus. Spheres were found more attractive to both female and male tephritid fruit flies than cubes, cylinders and rectangles of different surface on coffee (Nakagawa *et al.,* 1978). The most effective tactics developed for detecting the presence of adult *R. mendax* include baited Pherocon AM yellow sticky boards on blue berries (Prokopy and Coli 1978).

Robacker (1992) opined that female Mexican fruit fly, *A. ludens* preferred large spheres over large rectangles and small rectangles over small spheres in grapefruit orchard. Heath *et al.* (1995) found that McPhail traps with standard protein bait caught more Mexican fruit flies than either of the plastic traps at any doses of synthetic baits whereas Uchida *et al.* (1996) found that Jackson trap, orange and yellow bucket traps were more suitable for monitoring purpose as they captured more number of *C. capitata.*

Heath *et al.* (1996) reported that an open-bottom trap made of green opaque with a sticky insert captured more *C. capitata* than closed-bottom painted dry trap with a toxicant panel on sweet orange. According to Cornelius *et al.* (1999), greater number of Oriental fruit fly females were attracted to yellow coloured spheres and rectangular blocks of equivalent surface in guava orchards. Liburd *et al.* (2000) demonstrated that baited 9 cm diameter sphere was more effective in capturing *R. mendax* whereas yellow sticky boards captured significantly more fruit flies than sticky yellow Pherocon AM boards on blue berries in New Jersey. Mayer *et al.* (2000) obtained more trap catch of cherry fruit fly, *R. indifferens* on 10 cm red spheres followed by 8 and 12 cm compared to vertical and V oriented yellow rectangles.

Robacker and Heath (2001) reported that a sticky trap for fruit flies made from fruit fly adhesive paper (FFAP) covered with a plastic mesh of either 1.5×1.5 or 2.2×2.2 cm mesh size was as effective as Pherocon AM traps in capturing Mexican fruit fly, *A. ludens* on citrus orchard in Weslaco, Texas. Smith *et al.* (2001) found that Pherotech *Rhagoletis* traps were more efficient than the conventional Pherocon AM traps for monitoring *R. pomonella* in apple orchards whereas Stonehouse *et al.* (2002a) reported that square and oblong blocks were more effective in attracting *Bactrocera* spp. than round and hexagonal blocks.

In various trap designs, IIHR and open pan trap attracted significantly more number of *Bactrocera* spp. (Madhura and Viraktamath 2003). In guava, *B. correcta* was attracted to spheres and cylinders while *B. zonata* to bottle traps. However, *B. dorsalis* did not show any preference to sphere shaped trap (Rajitha and Viraktamath 2005a). But in mango ecosystem, *B. correcta* and *B. zonata* showed preference to spheres and bottles (Rajitha and Viraktamath 2005b). Eliopoulos (2007) reported that Glass-Plastic Elkofon trap attracted more *B. oleae* flies than any other five types of traps. Satisfactory catches were also given by the glass McPhail trap, plastic McPhail trap and plastic Elkofon trap, whereas low attractiveness was demonstrated by bottle trap and pouch trap in olive orchards in Greece. It is clear from findings of this study that trap captures of the olive fruit fly are significantly influenced by trap design. Vargas *et al.* (2010a) compared Jackson trap (standard Florida detection trap) and a Hawaii AWPM trap (standard Hawaii AWPM trap) having Farma Tech-Mallet-methyl eugenol or standard cue-lure trap with DDVP to trap the population of *B. dorsalis* and *B. cucurbitae,* respectively and found that both performed significantly different.

Chandaragi *et al.* (2012) reported that bottle trap was found to have significantly higher trap catch (41.13 fruit flies/trap/week) followed by cylinder trap (29.84 flies/trap/week) when five traps with different designs (bottle trap, cylinder, sphere, PCI trap and open trap) were used to capture fruit flies in mango orchards in Karnataka. Bjelis *et al.* (2014) reported that the visual-cue trap type Chromotrap (yellow sticky trap) in combination with ammonium bicarbonate was more effective tool for early detection, monitoring and mass trapping of cherry fruit fly, *R. cingulata* (Loew) than the other tested combinations of traps i.e. Tephri traps and modified McPhail traps in cherry orchards in northeast Slovenia.

Lasa *et al.* (2014) showed that under caged conditions, on mango, a commercial hemispherical trap with lateral holes (Maxitrap Plus) proved more attractive to *A. ludens* (both sexes) than five other commercial traps that were all baited with hydrolyzed protein. Daniel *et al.* (2014) conducted an experiment for the development of cost-efficient, lead chromate-free and more eco-friendly trap for monitoring and mass trapping of *R. cerasi* in cherry. Five different-coloured yellow panels and three different trap shapes i.e. cross, cube and cylinder were compared to a standard Rebell® amarillo trap. Rizk *et al.* (2014) conducted a study in peach orchards in Arab-Elmadabegh on four types of traps: bottle trap, glass McPhail trap, plastic McPhail trap and Abdel-Kawi trap baited with different doses of methyl eugenol. The results indicated that Abdel-Kawi trap charged with 0.5 ml methyl eugenol was the most effective trap.

Conclusions

Despite extensive research, fruit flies still remain a major threat to fruit and vegetable production in India. The damage and economic impact of fruit flies should be of great concern to all stakeholders along the fruit value chain. Smallholder farmers could be suffering from higher losses due to fruit fly infestation. The export potential of fresh fruits and vegetables could also be more threatened by these quarantine pests. A concerted effort is required by the fruit fly research communities to provide technologies, build capacity and create awareness on the importance of these pests for improving the horticulture industry. From the above description, it is clear that methyl eugenol based traps offer one of the most effective method of mass trapping of fruit flies. Since adult fruit flies use visual and olfactory stimuli to locate hosts, traps that combine visual and olfactory cue proved to be most efficient for capturing fruit flies. Although, a lot of work has been done on the development of various traps for the management, however, so far, no universal, effective trap has been developed to eradicate this pest. There is the need to establish an integrated approach including cultural practices such as collection and deep burying of infested and fallen fruits, tillage around the tree and the use of lures and traps as well as baits.

References

Abdullah, K., Latif, A., Khan, S. M. and Khan, M. A. 2007. Field test of the bait spray on periphery of host plants for the control of the fruit fly, *Myiopardalis pardalina* Bigot (Tephritidae: Diptera). *Pak Ent* 29: 91-94.

Agarwal, M. L. and Kumar, P. 1999. Relative efficacy of bait and attractant combinations against peach fruit fly, *Bactrocera zonata* (Saunders). *Pestology* 23: 23-26.

Allwood, A. J. 1997. Biology and Ecology: Prerequisites for understanding and managing Fruit Flies (Diptera:Tephritidae). *Management of Fruit Flies in the Pacific.* A Regional Symposium, Nadi, Fiji, 28-31 October 1996. *ACIAR Proceeding,* 76: 95-101.

Alyokhin, A. V., Messing, R. H. and Duan, J. J. 2000. Visual and olfactory stimuli and fruit maturity affect trap captures of oriental fruit flies (Diptera: Tephritidae). *Ecol Behav* 93: 644-49.

Anonymous. 2004. *Package of Practices for Horticultural Crops*, 470 pp. University of Agricultural Sciences, Dharwad.

Balasubramanian, G. E., Abraham, V., Vijayaraghavan, S., Subramaniam, T. R., Santhanaraman, T. and Gunasekaran, C. R. 1972. Use of male annihilation technique in the control of the Oriental fruit fly, *Dacus dorsalis* Hendel. *Indian J agric Sci* 42: 975-77.

Bateman, M. A. 1972. The ecology of fruit flies. *A Rev Ent* 17: 493-518.

Bateman, M. A. 1991. *The impact of fruit flies (fam. Tephritidae) on Australian Horticulture*. Horticultural Policy Council Report No. 3. ISBN 0 642 16110 0.

Belavadi, V. V. 1979. *Bionomics of the Oriental fruit fly, Dacus dorsalis Hendel (Diptera: Tephritidae) on guava (Psidium guajava L.) and its control by male annihilation.* Thesis, M. Sc. UAS-Bangalore.

Bhagat, D., Samanta, S. K. and Bhattacharya, S. 2013. Efficient management of fruit pests by pheromone nanogels. *Sci Rep* 3: 1-8.

Bindra, O. S. and Mann, G. S. 1978. An investigation into lure and lure traps for the guava fruit fly. *Indian J Hort* 35: 401-05.

Bjelis, M., Radunic, D., Miklave, J. and Seljak, G. 2014. Evaluation of trap types and food attractants for north America Cherry fruit fly-*Rhagoletis cingulata* Loew (Diptera: Tephritidae). pp 182. 9th *Int Symp Fruit Flies Econ Imp.* Bangkok, Thailand, 12-16 May, 2014.

Boller, E. F. 1969. Neuw uber die kirschenfliege: Frieiland veruche im Jahre 1969. *Schweizerische Zeitschrift Fur Obst Und Weinbau* 105: 566-72.

Boller, E. F. 1983. Biotechnical methods for the management of fruit fly populations. pp 342-52. In: Cavalloro R (ed) *Fruit Flies of Economic Importance*. CEC/IOBC Symposium, Athens, 1982. Balkema, Rotterdam.

Bose, P. C., Tiwari, L. D. and Mehrotra, K. N. 1979. Preliminary studies on the control of fruit fly in guava orchard by insecticide baits. *Indian J Ent* 41: 388-90.

Butani, D. K. 1979. *Insects and Fruits.* Periodicals Exports Book Agency, New Delhi, 415 pp.

Casagrande, E. 2010. Magnet MED system (attract and kill) for the control of the Mediterranean fruit fly (*Ceratitis capitata*). *Levante Agricola* 49: 303-06.

Chandaragi, M., Sugandhi, R., Vinay Kumar, M. M. and Uppar, V. 2012. Evaluation of different trap designs for capture of fruit flies in mango orchard. pp 28. In: Abstracts. *Int Conf Ent.* Feb 17-19, 2012. Punjabi University, Patiala.

Chiu, H. T. and Chu, Y. I. 1988. The male annihilation of Oriental fruit fly on Lamby Island. *China J Ent* 8: 81-94.

Christenson, L. E. and Foote, R. E. 1960. Biology of fruit flies. *A Rev Ent* 5: 171-92.

Chu, Y. I., Yeh, W. I. and Lu, C. Y. 1985. The development of poisoned dispenser for the control of Oriental fruit fly (*Dacus dorsalis* Hendel). *Pl Prot Bull Taiwan* 27: 413-21.

Clarke, A. R., Armstrong, K. F., Carmichel, A. E., Milne, J. R., Raghu, S., Roderick, G. K. and Yeates, D. K. 2005. Invasive phytophagous pests arising through a recent tropical evolutionary radiation: the *Bactrocera dorsalis* complex of fruit flies. *A Rev Ent* 50: 293-319.

Cornelius, M. L., Duan, J. J. and Messing, R. H. 1999. Visual stimuli and the response of female oriental fruit flies (Diptera: Tephritidae) to fruit-mimicking traps. *J econ Ent* 92: 121-29.

Cunningham, R. T. 1989. Male annihilation. In: Robinson A S and Hooper G (ed) *Fruit Flies: Their Biology, Natural Enemies and Control. World Crop Pests.* 3A. pp 345-51. Elsevier Science Publications, Amsterdam, The Netherlands.

Cunningham, R. T., Chambers, D. L. and Forbes, A. G. 1975. Oriental fruit fly: Thickened formulation of methyl eugenol in spot application for male annihilation. *J Econ Ent* 68: 861-63.

Cytrynowicz, M., Morgante, J. S. and De souza, H. M. L. 1982. Visual responses of South American fruit flies, *Anastrepha fraterculus,* and Mediterranean fruit flies, *Ceratitis capitata,* to colored rectangles and spheres. *Env Ent* 11: 1202-10.

Daniel, C., Mathis, S. and Feichtinger, G. 2014. A new visual trap for *Rhagoletis cerasi* (L.) (Diptera: Tephritidae). *Insects* 5: 564-76.

David, K. J. and Ramani, S. 2011. An illustrated key to fruit flies (Diptera: Tephritidae) from peninsular India and Andaman and Nicobar Islands. *Zootaxa* 3021: 1-31.

Drew, R. A. I. 1974. The response of fruit fly species (Diptera: Tephritidae) in South Pacific Asia area to male attractants. *J Aust Ent Soc* 13: 267-70.

Drew, R. A. I. and Hancock, D. L. 1994. The *Bactrocera dorsalis* complex of fruit flies (Diptera: Tephritidae: Dacinae) in Asia. *Bull Ent Res* 2 (Supple): 1-68.

Drew, R. A. I. and Hooper, G. H. S. 1981. The response of fruit fly species (Diptera: Tephritidae) in Australia to various attractants. *J Aust Ent Soc* 20: 201-05.

Eliopoulos, P. A. 2007. Evaluation of commercial traps of various designs for capturing the olive fruit fly *Bactrocera oleae* (Diptera: Tephritidae). *Int J Pest Mgmt* 53: 245-252.

Epsky, N. D., Heath, R. R., Guzman, A. and Meyer, W. L. 1995. Visual cue and chemical cue interactions in a dry trap with food-based synthetic attractant for *Ceratitis capitata* and *Anastrepha ludens* (Diptera: Tephritidae). *Env Ent* 24: 1387-95.

Ferrar, P. 2010. Fruit flies in Asia (especially Southeast Asia). Species, biology and management. Available online on www.scribd.com/doc/.../Fruit-Flies-in-Asia-Paper-paul-27-Aug-2010.

Fletcher, B. S. 1989. Life history strategies of tephritid fruit flies. In: Robinson A S and Hooper G (ed) *Fruit Flies: Their Biology, Natural Enemies and Control. World crop pests.* 3B. Elsevier Sci Publication, Amsterdam, pp 195-208.

Froerer, K. M., Peck, S. L., McQuate, G. T., Vargas, R. I., Jang, E. B. and McInnis, D. O. 2010. Long-distance movement of *Bactrocera dorsalis* (Diptera: Tephritidae) in Puna, Hawaii: How far can they go? *Am Ent* 56: 88-95.

Greany, P. D. and Szentes, I. A. 1979. Oviposition behaviour of laboratory reared and wild Caribbean fruit flies *Anastrepha suspensa* (Diptera: Tephritidae) : II selected physical influences. *Ent expl Appl* 26: 239-44.

Greany, P. D., Agee, H. R., Burditt, A. K. and Chambers, D. L. 1977. Field studies on colour preference of the Caribbean fruit fly, *Anastrepha suspense* (Diptera: Tephritidae), by use of fluorescent colours. *Ent expl Appl* 21: 63-70.

Greany, P. D., Burditt, A. K., Agee, H. R. and Chambers, D. L. 1978. Increasing effectiveness of visual traps for the Caribbean fruit fly, *Anastrepha suspense* (Diptera: Tephritidae), by use of fluorescent colors. *Ent expl Appl* 23: 20-25.

Han, P., Wang, X., Niu, C. Y., Dong, Y. C. and Zhu, J. Q. 2011. Population dynamics, phenology, and overwintering of *Bactrocera dorsalis* (Diptera: Tephritidae) in Hubei Province, China. *J Pest Sci* 84: 289-95.

Haq, I., Vreysen, M. J. B., Caceres, C., Shelly, T. E. and Hendrichs, J. 2014. Methyl eugenol aromatherapy enhances the mating competitiveness of male *Bactrocera carambolae* Drew & Hancock (Diptera: Tephritidae). *J Insect Physiol* 68: 1-6.

Hardy, D. E. 1979. Review of the economic fruit flies of the south pacific region. *Pac insects* 20: 429-32.

Heath, R. R., Epsky, N. D., Dueben, B. D. and Meyer, W. L. 1996. Systems to monitor and suppress Mediterranean fruit fly (Diptera: Tephritidae) populations. *Fl Ent* 79: 144-53.

Heath, R. R., Epsky, N. D., Guzman, A., Dueben, B. D., Manukian, A. and Meyer, W. L. 1995. Development of a dry plastic insect trap with food based synthetic attractant for the Mediterranean and the Mexican fruit flies (Diptera: Tephritidae). *J econ Ent* 88: 1307-15.

Howlett, F. M. 1912. The effects of oil of citronella on two species of *Dacus*. *Trans R Entomol Soc London* 60: 412-18.

Iwahashi, O. 1984. The control project of the Oriental fruit fly in Okinawa. *Chin J Ent* 4: 107-20.

Jalaluddin, S. M., Natarajan, K., Sadakathulla, S. and Rajukkannu, K. 1998. Effect of colour, height and dispenser on catch of guava fruit fly. In Reddy P P, Kumar N K K, Verghese A (ed): *Proc Symp: Advances IPM Hort Crops*. Association for the Advancement of Pest Management in Horticultual Ecosystems, Bangalore, pp 34-39.

Jamwal, V. S., Panwar, V., Sharma, A. and Prasad, C. S. 2015. Relative trapping efficiency of different colours in lure charged traps for two species of mango fruit flies in Saharanpur region. *4th Cong Insect Sci*. PAU, Ludhiana, 16-17 April, 2015.

Jhala, R. C., Sisodiya, D. B., Verghese, A., Mumford, J. D. and Stonehouse, J. M. 2005. Effect of colour on the attraction of parapheromone lures to tephritid fruit flies (*Bactrocera* spp.). *Pest Mgmt Hort Eco* 11: 170-71.

Jiji, T., Thomas, J., Singh, H. S., Jhala, R. C., Patel, R. K., Napolean, A., Senthilkumar, Vidya, C. V., Mohantha, A., Joshi, B. K., Stonehouse, J. M., Verghese, A. and Mumford, J. D. 2005. Attraction of Indian fruit flies to coloured spheres. *Pest Mgmt Hort Ecosys* 11: 88-90.

Kapoor, V. C., Grewal, J. S. and Beri, A. S. 1987. Fruit fly traps in the population study of fruit flies. pp 461. *Proc 2nd Int Symp Fruit Flies.* Crete (Greece), 1986, Elsevier Science Publication Co USA.

Katsoyannos, B. I. and Kouloussis, N. A. 2001. Captures of the olive fruit fly *Bactrocera oleae* on spheres of different colours. *Ent expl Appl* 100: 165-72.

Koyama, J., Tureja, T. and Tanaka, K. 1984. Eradication of the Oriental fruit fly (Diptera: Tephritidae) from the Okinawa Islands by a male annihilation method. *J econ Ent* 77: 468-72.

Kumar, N. K. K. and Sharma, D. R. 2012. Fruit pests of citrus. pp 203-209. Lead lecture. In: Souvenir and Abstracts. *Nat Dialogue Citrus Impr, Prod Utiliz* (*Theme: Climate Resilient Citrus Production System*). February 27-29, 2012. National Research Centre for Citrus, Nagpur.

Kumar, P. 2011. *Field Exercise Guide on Fruit Flies Integrated Pest Management for Farmer's Field Schools and Training of Trainers Area-Wide Integrated Pest Management of Fruit flies in South and Southeast Asian Countries.* Asian Fruit Fly IPM Project, Bangkok, Thailand. 63 pp.

Lakshmanan, P. L., Balasubramaniam, G. and Subramaniam, T. R. 1973. Effect of methyl eugenol in the control of the Oriental fruit fly *Dacus dorsalis* Hendel on mango. *Madras Agric J* 60: 628-29.

Lasa, R., Velazquez, O. E., Ortega, R. and Acosta, E. 2014. Efficacy of commercial traps and food odor attractants for mass trapping of *Anastrepha ludens* (Diptera: Tephritidae). *J econ Ent* 107: 198-205.

Leblanc, L. and Putoa, R. 2000. *Fruit flies in French Polynesia and Pitcairn Islands*. Secretariat of the Pacific Community Advisory Leafl., Suva, Fiji.

Liburd, O. E., Alm, S. R., Casagrande, R. A. and Polavarapu, S. 1998. Effect of trap colour, bait, shape and orientation in attraction of blueberry maggot (Diptera: Tephritidae) flies. *J econ Ent* 91: 243-49.

Liburd, O. E., Polavarapu, S., Alm, S. R. and Casagrande, R. A. 2000. Effect of trap size, placement and age on captures of blueberry maggot flies (Diptera: Tephritidae). *J Econ Ent* 93: 1452-58.

Liu, Y. C. 1991. Development of attractants for controlling the melon fly *Dacus cucurbitae* Coquillet in Taiwan. *Chinese J Ent Special Publication* 1: 115-29.

Liu, Y. C. and Lin, J. S. 1992. The attractiveness of 10% MC to melon fly *Dacus cucurbitae* Coquillett. *Pl Prot Bull Taipei* 34: 307-15.

Madhura, H. S. 2001. Management of fruit flies (Diptera: Tephritidae) using physical and chemical attractants. *M.Sc.* (*Agri.*) *Thesis*, University of Agricultural Sciences, Bangalore, pp 80.

Madhura, H. S. and Viraktamath, C. A. 2003. Efficacy of different traps in attracting fruit flies (Diptera: Tephritidae). *Pest Mgmt Hort Ecosys* 9:153-54.

Makhmoor, H. D. and Singh, S. T. 1998. Effective concentration of methyl eugenol for the control of guava fruit fly, *Dacus dorsalis* Hendel in guava orchards. *Ann Pl Prot Sci* 6: 165-69.

Mann, G. S. 1980. Population fluctuation of *Dacus dorsalis* Hendal in peach, guava and mango orchards at Ludhiana (Punjab). pp 49. *9th Ann Conf Ent Soc India*, Madurai, 11-13 February, 1984.

Mann, G. S. 1986. Potency of different insecticides at various spray intervals with/without protein hydrolysate bait against *Dacus dorsalis* Hendel infesting guava in Punjab. *6th Int Cong Pestic Chem,* Ottawa, Ontario, Canada. August 10-17, 1986.

Mann, G. S. 1990. Bioecology and management of fruit flies. *Summer Institute on Key Pests of India, their bioecology with special reference to Integrated Pest Management.* Department of Entomology, Punjab Agricultural University, Ludhiana. June 6-15, 1990: pp 236-57.

Manrakhan, A., Grout, T. and Hattingh, V. 2009. Combating the African invader fly *Bactrocera invadens*. *SA Fruit J* 8: 57-61.

Marwat, N. K., Hussain, N. and Khan, A. 1992. Suppression of *Dacus* spp. by male annihilation in guava orchard. *Pak J Zool* 24: 82-84.

Mayer, D. F., Long, L. E., Smith, T. J., Olsen, J., Riedel, H., Heath, R. R., Leskey, T. C. and Prokopy, R. J. 2000. Attraction of adult *Rhagoletis indifferens* (Diptera: Tephritidae) to unbaited and odour baited red spheres and yellow rectangles. *J Econ Ent*, 93: 347-51.

Mehta, P. K., Sood, P. and Prabhakar, C. S. 2010. Palam Trap: A novel triumph in fruit fly suppression in Himachal Pradesh. *Ent Reporter* 1: 8-9.

Metcalf, R. L. 1990. Chemical ecology of dacine fruit flies (Diptera: Tephritidae). *Ann ent Soc Am* 83: 1017-30.

Nagaraj, K. S., Jaganath, S., Srikanth, L. G. and Husainnaik, M. 2014a. Attraction of fruit fly to different quantities of methyl eugenol in mango orchard. *Tr Biosci* 7: 1113-15.

Nagaraj, K. S., Jaganath, S., Basavarajeshwari, and Srikanth, L. G. 2014b. Efficacy of different coloured traps in capturing fruit flies in mango orchard. *Tr Biosci* 7: 968-70.

Nakagawa, S., Prokopy, R. J., Wong, T. T. Y., Ziegler, J. R., Mitchell, S. M., Urago, T. and Harris, E. J. 1978. Visual orientation of *Ceratitis capitata* flies to fruit models. *Ent expl Appl* 24: 193-98.

Ohno, S., Tamura, Y., Haraguchi, D., Matsuyama, T. and Kohama, T. 2009. Re-invasions by *Bactrocera dorsalis* complex (Diptera: Tephritidae) occurred after its eradication in Okinawa, Japan, and local differences found in the frequency and temporal patterns of invasions. *App Ent Zool* 44: 643-54.

Patel, R. K., Jhala, R. C., Joshi, B. K., Sisodiya, D. B., Verghese, A., Mumford, J. D. and Stonehouse, J. M. 2005a. Effectiveness of solvents for soaked–block annihilation of male fruit flies in Gujarat. *Pest Mgmt Hort Ecosys* 11: 123-25.

Patel, Z. P., Jhala, R. C., Jagadale, V. S., Sisodya, D. B., Stonehouse, J. M., Verghese, A. and Mumford, J. D. 2005b. Effectiveness of woods for soaked-block annihilation of male fruit flies in Gujarat. *Pest Mgmt Hort Ecosys* 11: 117-20.

Prokopy, R. J. 1973. Dark enamel sphere capture as many apple maggot flies as fluorescent spheres. *Env Ent* 2: 953-54.

Prokopy, R. J. 1977. Stimuli influencing trophic relation in Tephritidae. *Collaq Inst CNSR* 265: 305-06.

Prokopy, R. J. and Coli, W. M. 1978. Selective traps for monitoring *Rhagoletis mendax* flies. *Prot Eco* 1: 45-53.

Rajitha, A. R. and Viraktamath, S. 2005a. Efficacy of different types of traps in attracting fruit flies in guava orchard at Dharwad, Karnataka. *Pest Mgmt econ Zoo* 131: 111-20.

Rajitha, A. R. and Viraktamath, S. 2005b. Responce of fruit flies to different types of traps in mango orchard. *Pest Mgmt Hort Ecosys* 11: 15-25.

Rajitha, A. R. and Viraktamath, S. 2006. Response of fruit flies (Diptera: Tephritidae) to different coloured bottle traps in guava orchard. *Pest Mgmt Hort Ecosys* 12: 170-72.

Ravikumar, C. H. and Viraktamath, S. 2006. Influence of size and number of holes in bottle traps containing methyl eugenol on the capturing efficiency of fruit flies in guava and mango orchards. *Pest Mgmt Econ Zoo* 14: 115-19.

Ravikumar, P. and Viraktamath, S. 2007. Attraction of fruit flies to different colours of methyl eugenol traps in guava and mango orchards. *Karnataka J Agric sci* 20: 749-51.

Ravikumar. 2005. *Studies on fruit flies in guava and mango orchards with special reference to their management through mass trapping.* Thesis, M.Sc.UAS-Dharwad.

Reji Rani, O. P., Paul, A. and Jiji, T. 2012. Field evaluation of methyl eugenol trap for the management of Oriental fruit fly, *Bactrocera dorsalis* Hendel (Diptera: Tephritidae) infesting mango. *Pest Mgmt Hort Ecosys* 18: 19-23.

Renou, M. and Guerrero, A. 2000. Insect pheromones in olfaction research and semiochemical-based pest control strategies. *Ann Rev Ent* 45: 605–30.

Rizk, M. M. A., Abdel-Galil, F. A., Temerak, S. A. H. and Darwish, D. Y. A. 2014. Factors affecting the efficacy of trapping system to the peach fruit fly (PFF) males, *Bactrocera zonata* (Saunders) (Diptera: Tephritidae). *Archives Phytopathol Pl Prot* 47: 490-98.

Robacker, D. C. 1992. Effects of shape and size of coloured traps on attractiveness to irradiated, laboratory-strain Mexican fruit flies (Diptera: Tephritidae). *Flo Ent* 75: 230-41.

Robacker, D. C. and Heath, R. R. 2001. Easy to handle sticky trap for fruit flies (Diptera: Tephritidae). *Flo Ent* 84: 302-04.

Robacker, D. C., Moreno, D. S. and Wolfenbarger, D. A. 1990. Effects of trap colour, height and placement around trees on capture of Mexican fruit flies (Diptera: Tephritidae). *J Econ Ent* 83: 412-19.

Roomi, M. W., Abbas, T., Shah, A. H., Robina, S., Qureshi, A. A., Sain, S. S. and Nasir, K. A, 1993. Control of fruit flies (*Dacus* sp.) by attractants of plant origin. *Anzeiger fiir Schadlingskunde*, *Pflanzeschutz, Umwelschutz* 66: 155-57.

Ros, J. P. and Seris, E. 2010. Transparent versus yellow color in traps against *Bactrocera oleae* (Rossi). pp 287-97. *8th Int Symp Fruit Flies Econ Imp.* Valencia, Spain, 26 September-1 October, 2010.

Sarada, G., Maheswari, T. U. and Purushotham, K. 2001 Effect of trap colour, height and placement around trees in capture of mango fruit flies. *J Appl Zoo Res* 12: 108-10.

Schutze, M. K., Krosch, M. N., Armstrong, F. K., Chapman, T. A., Englezou, A., Chomic, A., Cameron, S. L., Hailstones, D. and Clarke, A. R. 2012. Population structure of *Bactrocera dorsalis, B. papayae* and *B. philippinensis* (Diptera: Tephritidae) in southeast Asia: Evidence for a single species hypothesis using mitochondrial DNA and wing-shape data. *BMC Evol Biol* 12:130.

Shanker, M., Rao, S. R. K., Umamheswari, T. and Reddy, K. D. 2010. Effect of number and size of holes of trap on capturing efficiency of *Bactrocera* spp. in mango. *Ann Pl Protec Sci* 18: 223-82.

Sharma, D. R., Arora, P. K. and Singh, S. 2005. Insect and mite pests of fruit crops in Punjab: A Survey Report. *J Pl Sci Res* 21: 109-15.

Sharma, D. R., Bal, J. S. and Chahill, B. S. 2008. Assessment of reaction of guava germplasm against fruit fly and castor capsule borer. *Indian J Hort* 65: 145-51.

Sharma, D. R., Singh, S. and Aulakh, P. S. 2011. *Management offruit flies in fruit crops.* Department of Horticulture, Punjab Agricultural university, Ludhiana (December).

Sharma, K. 2012. Fruit fly management at Indian Agricultural Research Institute for production of organic fruits with male annihilation technique. pp 82. Abstracts. *Int Conf Ent,* Feb 17-19, Punjabi University, Patiala. 2012.

Shell, T. E., Epsky, N., Jang, E. B., Reyes-Flores, J. and Vargas, R. (ed) 2014. *Trapping and the Detection, Control, and Regulation of Tephritid Fruit Flies-Lures, Area-Wide Programs, and Trade Implications.* Pp 638. Springer, The Netherland.

Shelly, T. E. and Dewire, A. M. 1994. Chemically mediated mating success in male Oriental fruit flies (Diptera:Tephritidae). *Ann ent Soc Am* 87: 375-82.

Shelly, T. E., Nishimoto, J., Diaz, A., Leathers, J., War, M., Shoemaker, R., Al-Zubaidy, M. and Joseph, D. 2010. Capture probability of released males of two *Bactrocera* species (Diptera: Tephritidae) in detection traps in California. *J econ Ent* 103: 2042-51.

Singh, Sandeep. 2006. Pest Management. In: Mehan, V. K., Rattanpal, H. S. and Singh, S. (ed) Annual Report (2005-2006). AICRP on Tropical Fruits. Department of Horticulture, Punjab Agricultural University, Ludhiana. 120 pp.

Singh, Sandeep. 2008a. Fruit fly infestation on Kinnow in Punjab. Pp 200-201. In: Joia, B. S., Sharma, D. R., Dilawari, V. K. and Pathania, P. C. (ed). Contributory Papers. *Proc of 2nd Congress on Insect Science: Pest Management in Global Context* (February 21-22, 2008). INSAIS, PAU, Ludhiana.

Singh, Sandeep. 2008b. Studies on host preferences of fruit flies, *Bactrocera dorsalis* (Hendel) and *Bactrocera zonata* Saunders in Punjab, India. pp 249. 11th International Citrus Congress, Wuhan, China, 26-30 October, 2008.

Singh, Sandeep. 2008c. Survey and surveillance of insect pests and natural enemies of citrus in Punjab. pp 132-33. Souvenir and Abstract. *Nat Symp Citriculture: Emerging Trends.* National Research Centre for Citrus, Nagpur, July 24-26, 2008.

Singh, Sandeep. 2012. *Development and Management of fruit flies, Bactrocera* spp. *on different fruit crops.* Ph.D.Dissertation, PAU, Ludhiana.

Singh, S. and Sharma, D. R. 2011. Comparison of the trapping efficacy of different types of methyl eugenol based traps against fruit flies, *Bactrocera* spp. infesting Kinnow mandarin in the Indian Punjab. *J Insect Sci* 24: 109-14.

Singh, Sandeep. and Sharma, D. R. 2012. Abundance and management of fruit flies on peach through male annihilation technique (MAT) using methyl eugenol based mineral water bottle traps. *J Insect Sci* 25: 135-43.

Singh, Sandeep. and Sharma, D. R. 2013. Management of fruit flies in rainy season guava through male annihilation technique using methyl eugenol based traps. *Indian J Hort* 70: 512-18.

Singh, Sandeep., Dhillon, W. S. and Sharma, D. R. 2011. Methyl Eugenol based male annihilation technique (MAT): An eco-friendly approach for monitoring and trapping fruit flies, *Bactrocera* spp. on Pear in Punjab: A step towards organic farming. pp 223-26. In: Dhillon, W. S., Aulakh, P. S., Singh, H., Gill, P. P. S. and Singh, N. (ed) Climate Change and Fruit production. *Proc Nat Sem Impact Climate Change Fruit Crops* (*ICCFC-2010*). 6th to 8th Oct. 2010, Department of Horticulture, Punjab Agricultural University, Ludhiana.

Singh, Sandeep., Sharma, D. R. and Kular, J. S. 2015. Eco-friendly management of fruit flies, *Bactrocera* spp. in peach with methyl eugenol based traps in Punjab. *Agric Res J* 52: 47-49.

Singh, Sandeep., Sharma, D. R., Kular, J. S., Gill, M. I. S., Arora, N. K., Bons, M. S., Singh, B., Boora, R. S., Kaur, A., Saini, M. K., Pandha, Y. S., Chahal, T. S., Kumar, G., Singh, B., Singh, S., Pandher, S., Sharma, R. K. and Kaur, P. 2014a. Eco-friendly management of fruit flies, *Bactrocera* spp. in guava with methyl eugenol based traps in Punjab. *Indian J Ecol* 41: 365-67.

Singh, Sandeep., Sharma, D. R., Kular, J. S., Singh, P., Gill, P. P. S., Singh, N., Bons, M. S., Singh, B., Kaur, A., Saini, M. K., Singh, B., Pandha, Y. S., Chahal, T. S., Kumar, G. and Kaur, P. 2014b. Eco-friendly management of fruit flies, *Bactrocera* spp. in pear with methyl eugenol based traps at different locations in Punjab. *J Insect Sci* 27: 57-62.

Singh, S. P. 1993. Integrated pest management in horticultural crops. *Indian Hort* 38: 25-40.

Singh, S. S., Yadav, S. K., Mayank, K. R. and Singh, V. B. 2009. Effect of new Eco-trap, methyl eugenol bait and carbaryl against fruit fly, *Bactrocera dorsalis* (Hendel) in mango. *Indian J Ent* 71: 236-39.

Singh, U. B., Singh, H. M. and Singh, A. K. 2007. Seasonal occurrence of fruit flies in Eastern Uttar Pradesh. *J appl Zool Res* 18: 124-27.

Sivinski, J. 1990. Coloured spherical traps for the capture of Caribbean fruit fly, *Anastrepha suspense. Flo Ent* 73: 123-28.

Smith, R. F., Mazerolle, M., Estabrooks, E. and Vincent, C. 2001. Monitoring apple maggot, *Rhagoletis pomonella* (Walsh) (Diptera: Tephritidae) populations in commercial apple orchards of Nova Scotia, New Brunswick and Quebec. *Technical Report, Atlantic Food and Horticulture Research Centre*, pp 1-12.

Stark, J. D. and Vargas, R. I. 1992. Differential response of male oriental fruit fly (Diptera: Tephritidae) to coloured traps baited with methyl eugenol. *J econ Ent* 85: 802-12.

Steiner, L. F., Mitchell, W. C., Harris, E. J., Kozuma, T. T. and Fujimoto, M. S. 1965. Oriental fruit fly eradication by male annihilation. *J econ Ent* 53: 961-64.

Stonehouse, J. M. 2001. An overview of fruit fly research knowledge and needs in the Indian Ocean region. In: *Pro Second Nat Symp Integ Pest Mgmt (IPM) in Hort Crops: New Molecules Biopesticides Environment*, Bangalore, pp 21-23.

Stonehouse, J. M., Afzal, A., Zia, Q., Mumford, J., Poswal, A. and Mahmood, R. 2002a. "Single-killing-point" field assessment of bait and lure control of fruit flies (Diptera: Tephritidae) in Pakistan. *Crop Prot* 21: 651-59.

Stonehouse, J. M., Mahmood, R., Poswal, A., Mumford, J., Baloch, K. N., Chaudhary, Z. M., Makhdum, A. H., Mustafa, G. and Huggett, D. 2002b. Farm field assessments of fruit flies (Diptera: Tephritidae) in Pakistan: distribution, damage and control. *Crop Prot* 21: 661-69.

Stonehouse, J. M., Verghese, A., Mumford, J. D., Thomas, J., Jijli, T., Faleiro, R., Patel, Z. P., Jhala, R. C., Patel, R. K., Shukla, R. P., Satpathy. S., Singh, H. S., Singh, A. and Sardana, H. R. 2005. Research conclusions and recommendations for the on-farm IPM of tephritid fruit flies in India. *Pest Mgmt Hort Ecosys* 11: 172-180.

Sureshbabu, K. and Viraktamath, S. 2003. Species diversity and population dynamics of fruit flies (Diptera: Tephritidae) on mango in Northern Karnataka. *Pest Mgmt econ Zoo* 11: 103-10.

Uchida, G. K., Walsh, W. A., Encarnacion, C., Vargas, R. I., Stark, J. D., Beardsley, J. W. and Mcinnis, D. O. 1996. Design and relative efficiency of Mediterranean fruit fly (Diptera: Tephritidae) bucket traps. *J econ Ent* 89: 1137-42.

Ushio, S., Kengo, Y., Kazutoshi, N. and Keize, W. 1982. Eradication of the Oriental fruit fly from Amami islands by male annihilation (Diptera:Tephritidae). *Japan J Appl Ent Zool.* 26: 1-9.

Vargas, R. I., Burns, R. E., Mau, R. F. L., Stark, J. D., Cook, P. and Pinero, J. C. 2009a. Captures in methyl eugenol and cue-lure detection traps with and without insecticides and with a farma tech solid lure and insecticide dispenser. *J Econ Ent* 102: 552-57.

Vargas, R. I., Mau, R. F. L., Jang, E. B., Faust, R. M. and Wong, L. 2008a. The Hawaii fruit fly area wide pest management programme. pp 300-25. In: Koul O, Cuperus G W and Elliott (ed) *Area wide Pest Management: Theory and Implementation*. CAB International.

Vargas, R. I., Mau, R. F. L., Stark, J. D., Pinero, J. C., Leblanc, L. and Souder, S. K. 2010a. Evaluation of methyl eugenol and cue-lure traps with solid lure and insecticide dispensers for fruit fly monitoring and male annihilation in the Hawaii area wide pest management program. *J econ Ent* 103: 409-15.

Vargas, R. I., Miyashita, D. and Nishida, T. 1984. Life-history and demographic parameters of three laboratory reared tephritids (Diptera: Tephritidae). *Ann Ent Soc Am* 77: 651-56.

Vargas, R. I., Pinero, J. C., Mau, R. F. L., Jang, E. B., Klungness, L. M., McInnis, D. O., Harris, E. B., McQuate, G. T., Bautista, R. C. and Wong, L. 2010b. Area-wide suppression of the Mediterranean fruit fly, *Ceratitis capitata*, and the Oriental fruit fly, *Bactrocera dorsalis*, in Kamuela, Hawaii. *J Insect Sci* 10:135 available online: insectscience.org/10.135.

Vargas, R. I., Pinero, J. C., Mau, R. F. L., Stark, J. D., Hertlein, M., Mafra-Neto, A., Coler, R. and Getchell, A. 2009b. Attraction and mortality of oriental fruit flies (Diptera:Tephritidae) to SPLAT-MAT methyl eugenol with spinosad. *Ent expl Appl* 131: 286-93.

Vargas, R. I., Stark, J. D., Hertlein, M., Mafra-Neto, A., Coler, R. and Pinero, J. C. 2008b. Evaluation of SPLAT with spinosad and methyl eugenol or cue-lure for "attract-and-kill" of Oriental and melon fruit flies (Diptera:Tephritidae) in Hawaii. *J econ Ent* 101: 759-68.

Vargas, R. I., Stark, J. D., Prokopy, R. J. and Green, T. A. 1991. Response of oriental fruit fly (Diptera: Tephritidae) and associated parasitoids to different colour spheres. *J econ Ent* 84: 1503-07.

Verghese, A. 1998. Methyl eugenol attracts female mango fruit fly, *Bactrocera dorsalis* Hendel. *Insect env* 4: 101.

Verghese, A. and Mumford, J. D. 2010. Developing and implementing area-wide integrated management of mango fruit fly, *Bactrocera dorsalis* (Hendel) in South India. pp 189-99. *8th Int Symp Fruit Flies Econ Imp.* Valencia, Spain, 26 September-1 October, 2010.

Verghese, A., Shinananda, T. N. and Hegde, M. R. 2012. Status and area-wide integrated management of mango fruit fly, *Bactrocera dorsalis* (Hendel) in South India. Lead paper. In: Ameta O P, Swaminathan R, Sharma U S and Bajpai N K (ed) *Nat Sem Emerging Pest Problems Bio-rational Mgmt.* 2-3 March, 2012, Udaipur.

Verghese, A., Sreedevi, K., Nagaraju, D. K. and Jayanthi Mala, B. R. 2006. A farmer-friendly trap for the management of the fruit fly *Bactrocera* spp. (Tephritidae: Diptera). *Pest Mgmt Hort Ecosys* 12: 164-67.

Verghese, A., Tandon, P. L. and Stonehouse, J. M. 2004. Economic evaluation of the integrated management of the Oriental fruit fly *Bactrocera dorsalis* (Diptera: Tephritidae) in mango in India. *Crop Prot* 23: 61-63.

Verma, J. S. and Nath, A. 2006. Management of fruit flies through trapping-a review. *Agric Rev* 27: 44-52.

Viraktamath, S. and Ravikumar, C. 2006. Management of fruit flies through mass trapping in guava at Dharwad. *Pest Mgmt Hort Ecosys* 12: 137-42.

White, I. M. and Elson-Harris, M. M. 1992. *Fruit Flies of Economic Significance: Their Identification and Bionomics*. CAB International, Wallingford, 601 pp.

Yee, W. L. 2011. Evaluation of yellow rectangle traps coated with hot melt pressure sensitive adhesive and sticky gel against *Rhagoletis indifferens* (Diptera: Tephritidae). *Journal of Economic Entomology* 104: 909-19.

Yee, W. L. 2013. Captures of *Rhagoletis indifferens* (Diptera:Tephritidae) and non-target insects on red spheres versus yellow spheres and panels. *Journal of Economic Entomology* 106: 2109-17.

Yee, W. L. 2015. Commercial yellow sticky strips more attractive than yellow boards to western cherry fruit fly (Diptera: Tephritidae). *Journal of Applied Entomology* 139: 289-301.

10
Organic Pest Management for Biodynamic Farming

B. L. Jakhar

Inorganic pest management systems using external inputs such as chemical fertilizers, pesticides, high yielding varieties have resulted in increased yields in certain areas, but resulted in ecological degradation hence a decrease in productivity. Under organic farming systems, the fundamental components and natural processes of ecosystems, such as soil organism activities, nutrient cycling, and species distribution and competition are used to work directly and indirectly as farm management tools. For example, crops are rotated, planting and harvesting dates are carefully planned, and habitat that supplies resources for beneficial organisms are provided. Soil fertility and crop nutrients are managed through tillage and cultivation practices, crop rotations, cover crops, and supplemented with manure, composts, crop waste material and allowed substances.

Bio intensive IPM incorporates ecological and economic factors into agricultural system design and decision making and addresses public concerns about environmental quality and food safety. The benefits of implementing bio intensive IPM can include reduced chemical input costs, reduced on-farm and off-farm environmental impacts, and more effective and sustainable pest management. An ecology-based IPM has the potential of decreasing inputs of fuel, machinery, and synthetic chemicals—all of which are energy intensive and increasingly costly in terms of financial and environmental impact. Such reductions will benefit the grower and society.

A. Cultural practices

Cultural control is manipulations of the agro ecosystem that make the cropping system less friendly to the establishment and proliferation of pest population. Although they are designed to have positive effects on farm ecology and pest management, negative impacts may also result, due to variations in weather or changes in crop management.

Clean sanitation

Sanitation involves removing and destroying the over wintering or breeding sites of the pest as well as preventing a new pest from establishing on the farm (e.g., not allowing off-farm soil from farm equipment to spread nematodes or plant pathogens to your land). This strategy has been particularly useful in horticultural and tree-fruit crop situations involving twig and branch pests. If, however, sanitation involves removal of crop residues from the soil surface, the soil is left exposed to erosion by wind and water. As with so many decisions in farming, both the short- and long-term benefits of each action should be considered when trade offs like this are involved. Field sanitation by removing the previous crop straw, proper drainage to avoid water logging, and regulation of plant density were found effective for the management of *Pythium;* stalk rot of maize. Stripping off lower 2-3 leaves along with sheath when the crop is 30-35 days old proved very effective for the management of banded and leaf and sheath blight disease. Removal of 'Kans' grass from adjoining area of the main crop field helped in the reduction of downy mildew incidence in maize.

Deep ploughing in soil during summer months

Insect pest possess their inactive stage in soil. It is therefore advisable to plough the soil during hot month. Pupa are exposed to sun heat and also destroyed by predator's birds. This practice is very useful for the control of white grub, hairy caterpillar, stem borer, termite and cutworms.

Plant spacing

Spacing of plants heavily influences the development of plant diseases and weed problems. The distance between plants and rows, the shape of beds, and the height of plants influence air flow across the crop, which in turn determines how long the leaves remain damp from rain and morning dew. Better air flow will decrease the incidence of plant disease. However, increased air flow through wider spacing will also allow more sunlight to the ground, which may increase weed problems. This is another instance in which detailed knowledge of the crop ecology is necessary to determine the best pest management strategies.

Adjustment in the date of sowing

Altered planting dates can at times be used to avoid specific insects, weeds, or diseases. For example, squash bug infestations on cucurbits can be decreased by the delayed planting strategy, i.e., waiting to establish the cucurbit crop until overwintering adult squash bugs have died. Changes in sowing and planting dates can help in the avoidance of egg laying period of certain pests, the establishment of tolerant plant before pest occurrences, the maturation of crop before abundance of pest. By adjusting the date of sowing, crop is escaped by the following pest attack.

- Cotton- last week of May-pest attack will be less
- Castor – 3rd week of august- Semi looper larvae is minimized
- Mustard- 15-20 October crop is escaped from aphid damage.

Mulches

Mulches, living or non-living, are useful for suppression of weeds, insect pests, and some plant diseases. Hay and straw, for example, provide habitat for spiders. Research in Tennessee showed a 70% reduction in damage to vegetables by insect pests when hay or straw was used as mulch. The difference was due to spiders, which find mulch more habitable than bare ground. Other researchers have found that living mulches of various clovers reduce insect pest damage to vegetables and orchard crops. Again, this reduction is due to natural predators and parasites provided habitat by the clovers. Vetch has been used as both a nitrogen source and as weed suppressive mulch in tomatoes in Maryland. Growers must be aware that mulching may also provide a more friendly environment for slugs and snails, which can be particularly damaging at the seedling stage. Mulching helps to minimize the spread of soil-borne plant pathogens by preventing their transmission through soil splash. Mulch, if heavy enough, prevents the germination of many annual weed seeds. Winged aphids are repelled by silver or aluminium coloured mulches. Recent spring time field tests at the Agricultural Research Service in Florence, South Carolina, have indicated that red plastic mulch suppresses root-knot nematode damage in tomatoes by diverting resources away from the roots and nematodes towards the foliage and fruit.

Crop rotation

Crop rotation radically alters the environment both above and below ground, usually to the disadvantage of pests of the previous crop. The same crop grown year after year on the same field will inevitably build up populations of organisms that feed on that plant or in the case of weeds, have a life cycle similar to that of the crop. Add to this the disruptive effect of pesticides on species diversity, both above and below ground, and the result is an unstable system in which slight stresses (e.g., new pest variety or drought) can devastate the crop.

An enforced rotation program in the Imperial Valley of California has effectively controlled the sugar beet cyst nematode. Under this program, sugar beets may not be grown more than two years in a row or more than four years out of ten in clean fields (i.e., non-infested fields). In infested fields, every year of a sugar beet crop must be followed by three years of a non-host crop. Other nematode pests commonly controlled with crop rotation methods include the golden nematode of potato, many root-knot nematodes, and the soybean cyst nematode. Incidence was also reduced in the cropping system and in plots applied with farm yard manure (19.20 %) or neem cake @ 200kg/ac. (16.40%) compared to control (25.20%) (Jayaraj *et.al.*,1991).

Trap crop

Pests can be strongly attracted by certain plants. When these are sown in the field or along side it, insect will together on them and can thus be easily controlled. i.e. cotton bollworm *Helicoverpa armigera* prefer maize and lays eggs on cotton only when grown as a sole crop. When a few row of maize are sown in the cotton field the eggs will be laid on these plants and can be destroyed. Growing mustard as trap crop (two rows) in between cabbage or cauliflower, manages the diamond back moth incidence. First mustard crop is sown prior to cabbage planting, 20 days old cabbage or cauliflower seedlings are then planted. Growing castor along the border of cotton field and irrigation channel act as indicator or trap crop for *Spodoptera litura.* Planting 40 days old African marigold and 25 days old tomato seedling (1:16 rows) simultaneously reduce *H. armigera.*

Growing trap crop like marigold which attracted pest like American boll worm to lay eggs, barrier crop like maize, jowar to prevent migration of sucking pest like aphid and guard crops like castor which attract *S. litura* in cotton field was reported by Murthy and Venkateshwarulu (1998).

Other Cropping Structure Options

Multiple cropping is the sequential production of more than one crop on the same land in one year. Depending on the type of cropping sequence used, multiple cropping can be useful as a weed control measure, particularly when the second crop is inter planted into the first.

Interpolating is seeding or planting a crop into a growing stand, for example over seeding a cover crop into a grain stand. There may be microclimate advantages (e.g., timing, wind protection, and less radical temperature and humidity changes) as well as disadvantages (competition for light, water, nutrients) to this strategy. By keeping the soil covered, inter planting may also help protect soil against erosion from wind and rain.

Intercropping is the practice of growing two or more crops in the same, alternate, or paired rows in the same area. This technique is particularly appropriate in vegetable production. The advantage of intercropping is that the increased diversity helps "disguise" crops from insect pests, and if done well, may allow for more efficient utilization of limited soil and water resources. Disadvantages may relate to ease of managing two different crop species with potentially different nutrient, water, and light needs, and differences in harvesting time and method in close proximity to each other. Green gram intercropped sugarcane recorded 77 per cent decrease in sugarcane early shoot borer incidence over control (Krishnamurthi and Palanisamy, 1995).

Strip cropping is the practice of growing two or more crops in different strips across a field wide enough for independent cultivation (e.g., alternating six-row blocks of soybeans and corn or alternating strips of alfalfa and cotton or alfalfa and corn). It is commonly practiced to help reduce soil erosion in hilly areas. Like intercropping, strip cropping increases the diversity of a cropping area, which in turn may help "disguise" the crops from pests. Another advantage to this system is that one of the crops may act as a reservoir and/or food source for beneficial organisms. However, much more research is needed on the complex interactions between various paired crops and their pest/predator complexes.

The options described above can be integrated with no-till cultivation schemes and all its variations (strip till, ridge till, etc.) as well as with hedge rows and intercrops designed for beneficial organism habitat. With all the cropping and tillage options available, it is possible, with creative and informed management, to evolve a biologically diverse, pest-suppressive farming system appropriate to the unique environment of each farm.

B. Mechanical and Physical Control

Methods included in this category utilize some physical component of the environment, such as temperature, humidity, or light, to the detriment of the pest. Common examples are flaming, flooding, soil solarization, and plastic mulches to kill weeds or to prevent weed seed germination.

Heat or steam sterilization of soil is commonly used in greenhouse operations for control of soil-borne pests. Floating row covers over vegetable crops exclude flea beetles, cucumber beetles, and adults of the onion, carrot, cabbage, and seed corn root maggots. Insect screens are used in greenhouses to prevent aphids, thrips, mites, and other pests from entering ventilation ducts. Large, multi-row vacuum machines have been used for pest management in strawberries and vegetable crops. Cold storage reduces post-harvest disease problems on produce.

Although generally used in small or localized situations, some methods of mechanical/physical control are finding wider acceptance because they are generally more friendly to the environment.

- Mango stem borer - It is hides the whole month by the grub it can kill by wire.
- Leaves having egg masses / 1st instar gregarious larvae should be collected and destroyed.
- Big sized larvae of *Helicoverpa* / spodoptera / semi looper / weevil / beetle / hairy caterpillar should be collected and destroyed.

Use of light trap

It is one of the important components of IPM. We can know the appearance and the intensity of the pest. It is effective for phototropic insects. It is successfully used against hairy cater pillar, *Helicoverpa spodoptera litura*, diamondback moth, paddy stem borer, beetle of white grub, grasshopper, mole cricket. Light trap should be installed in the field and collect pest and they should be destroyed next day. In case of hairy caterpillar and white grub beetles light trap should be installed immediately after first rainfall.

Sex pheromone trap

Monitoring of pest population is a pre-requisite for efficient pest management strategy. Several methods are used for determining the pest population. Pheromone trap is one such tool, which is used for monitoring of pests. Sex pheromones are bio-chemicals released usually by female insects as a means of chemical communication to attract males for mating. They are highly specific and are perceived by the males of the same species only. The synthetic pheromones are now is used for luring a variety of lepidopterous pests in small entrapments popularly called pheromone traps. These traps are effectively used in cotton, rice, vegetable and many other crops. Two traps per acre are often sufficient for monitoring the pest population. Pheromones are also exploited for mass trapping of certain pests particularly yellow stem borer in rice, Usually ten traps per acre are used to mass trap the moths to reduce their population. Of late, pheromones are also used for disruption of mating cycle. Small bits of rope impregnated with pheromones are strewn for dispensing this chemical in the field. The crop environment pervaded by pheromone confuses the females to the extent that their mating get disrupted in spite of the fact that males are there in the field. The unmated females are rendered incapable of laying the eggs. Pheromones can be effectively used in organic farming. Lures are species specific hence lures are used for specific pest. Lure should be changed at 21 days interval. In market pheromone are available for some pests marketed as Spodolure lure, Heli lure, Ervit lure, Luci lure, DBM lure, Gossylure, Cue lure, Medi lure.

C. Pest resistant variety

Plant have their own sophisticated mechanism to protect them attack by pests. These range from physical deterrent to biochemical substance which either act as chemical signal in the ecosystem, sending message via the senses such as taste and smell to discourage herbivores activity or are directly toxic or may cause sterility or failure to reach sexual mortality. There should be a constant watch to update such genotype as under natural environment new pest are created and attack the hitherto resistant varieties. In resistant variety *viz.* MDU 3 (gall midge), PY 3, CO42 (BPH) should be used. To resist sorghum shoot fly incidence CSH 15R can be used. Ground nut resistant varieties like Robut 33-1, Kadiri 3, ICGS 86031 should be grown in endemic area to reduce the risk of thrips damage and bud necrosis. In case of cotton white fly tolerant varieties like JGJ 14545, LK 861, Supriya and Kanchan should be grown in endemic area. V-797 of cotton is also resistant to sucking pests. GC-4 variety of cumin is resistant to wilt. GCH-7 castor variety is found resistant to wilt.

D. Botanical products

Plants possess certain chemicals, which adversely affect the insect behavior. It attracts/ repel or some time they are toxic on their growth stages. Following plant materials are used against various insect pests.

- Tobacco decoction -
    - Soft bodied insect pests and leaf miner
- Neem based products -
    - Sucking pests, Hairy caterpillar, Helicoverpa, DBM, Grasshopper
- Pyrethrum – Housefly, cockroach, mosquito
    - Mustard aphid (leaf/seed powder) pulse beetle
- Mint leaves - pulse beetle
- Jatropha leaves – sucking pests

E. Biological control agents

Disease control

Eco-friendly organic farming technologies for plant protection have been gaining importance in recent years. Some of the plant diseases that can be controlled by antagonistic fungi and bacteria are as follows: *Aspergillus niger* AN 27 was found effective against several diseases caused by soil borne pathogens, *viz.* wilt *(Fusarium oxysporum, F. Solani)*, charcoal rot *(Macrophomina phaseolina)*, damping off (*Pythium aphanidermatum*). Sheath blight

(*Rhizoctonia solani*) and stalk rot (*Sclerotinia sclerotiorum*) by a single application under different agro-climates in cereals, millets, pulses, oilseeds, fruits, tuber vegetables, and ornamental, fodder and fiber crops. The commercial preparation is known by the name, Kalisena. Rice seeds treated with *Pseudomonas aeruginosa* and *P. putida* reduced sheath blight infection (*Rhizoctonia solani*) in rice by 65-72 per cent in comparison to untreated check. *P. fluorescens* was also found effective against banded leaf and sheath blight fungus (*R. Solani* f. sp. Sasakii). *Trichoderma harzianum* as fungal antagonist proved effective against *Macrophomina phaseolina* (charcoal rot) in several plant species. Application of rooting media, peat moss + wheat bran + perlite (2:1:2) mixed with *Trichoderma harzianum* (TH3) 14-16 days before transplanting of carnation was found effective and reduced 76-78% of wilt incidence caused by *F. oxysporum* F. sp. dianthi and increased the yield. It was followed by cocopeat + wheat bran + perlite (2:1:2). *T. harzianum* isolate TH3, was found most effective against *Colletotrichum capsici* causing fruit rot and die back of chilli in vitro followed by isolates TH4 and TH 1. Two sprays of conidial suspension of *T. harzianum* (TH3), one before 48 h of inoculation and second after 15 days of inoculation, were found effective in which lowest fruit rot intensity (44.5%) with 32.9% disease reduction was recorded. It was followed by sprays of *T. viride* – 2, and *Aspergillus niger* AN-27. Five *pseudomonads* were causing bacterial blight and *Sclerotium rolfsii* and *Rhizoctonia solani* causing damping off in cotton. Seed bacterisation with *Pseudomonas fluorescens* (PR 8) reduced damping off disease incidence caused by *R. solani and S. rolfisii* by 84 and 76% respectively. Seed bacterisation with the same isolate of *P. fluorescens* reduced the cotyledonary infection. *Trichoderma harzianum* and *Pseudomonas fluorescens* effectively suppressed mycelial growth, sclerotial production and germination of *Rhizoctonia solani* causing root rot of wheat. *Chaetomium globosum* Kunze Fr. has been identified as a potential biocontrol agent of spot blotch of wheat caused by *Drechslera sorokiniana*. It is also antagonistic to *Ascochyta rabiei* and *Fusarium oxysporium* ciceri causing ascochyta blight and wilt of chickpea respectively. The mechanism of action is through antibiosis. The antagonist is sprayed on the crop @ 106 ascospores/ml for the control of spot blotch of wheat and ascochyta blight of chickpea. For the control of chickpea wilt, the bioformulation is amended in the soil @ 8 g/1 meter row in the field before sowing. Many botanicals have the potential to control pests and diseases of plant. Extracts of neem, custard apple and callophyllum (undai) seed can control a wide range of insects, bacterial and fungi.

Insect pest control

Predators

Chrysoperla, ladybird beetle, syrphid fly are important predator of soft bodies insects Three releases of chrysoprela larvae can made at weekly interval for the control of soft-bodied insect pest of cotton bollworms @ 10,000 larvae/ ha. *Chilocorus nigrolotus is* effective predator of sugarcane scale insect. In sugarcane growing areas, grape vine mealy bug is important limiting factor. In grape arched for its management. *Cryptolaemus montrouzieri* is potent predator of grapevine mealy bug.

Parasite

Trichogramma spp. is effective egg parasite of most lepidopteran pests viz. cotton boll worm, rice stem borer, semi looper, diamond back moth. It is used @ 1.5 lac./ha/week. Six releases are made during cotton crop season. Similarly *Epiricania melanoleuca* is also found effective parasite of sugarcane pyrilla. It is used @ 2000 pupa or 2 lac. eggs /ha in pyrilla sugarcane infested field.

Microbial agents

Biopesticides' are certain types of pesticides derived from such natural materials as animals, plants, bacteria, and certain minerals. These include for example; fungi such as *Beauveria* sp., bacteria such as *Bacillus* sp. Biopesticides also play an important role in providing pest management tools in areas where pesticide resistance, niche markets and environmental concerns limit the use of chemical pesticide products. Biopesticides in general- (a) have a narrow target range and a very specific mode of action. (b) are slow acting. (c) have relatively critical application times. (d) suppress, rather than eliminate, a pest population. (e) have limited field persistence and a short shelf life. (f) are safer to humans and the environment than conventional pesticide. (g) present no residue problems. In market HNPV, SNPV are available for the management of *Helicoverpa* and *Spodoptera.* It is used during late evening. Bacterial preparations are used for leaf feeder pests of cabbage, cotton and ground nut. Fungal pathogen are used against soil pests and sucking pests also.

Organic manure

Organic manure in broad sence include compost from rural and urban waste, crop residues, agro industrtial bio waste and green manures, apart from commonly used FYM. Organic manure improves soil physical condition including soil porosity and water holding capacity and microbial environment, replenishes essential micronutrient in soil, increases the utilizations efficiencies of applied fertilizers and favors micronutrient availability to the plant. Organic manure is of paramount importance not only in augmenting the crop production but also for making the agriculture sustainable as an ecofriendly means of soil health management. It is well eatablished that FYM plays an additional role than its capacity to contribute NPK. Addition of organic manures such as FYM, recycling of organic waste through composting, green manures and biological inputs like vermicompost and biofertilisers etc. constitute important for plant nutrient management in organic farming, .similarly it also takes the utmost average of the natural mechanism for pest management with utilization of bioagent such as predators and parasites available in nature in plenty. Castor/ Neem/ Tobacco cake are useful against soil insect pest and nematode.

Conclusion

Chemical input in agriculture resulted in problems to ecosystem, thereby there is need to conserve biotic balance in the agroecosystem. Organic agriculture is promising and sustainable. However complete dependence on organic farming is risky, since achieving targeted yield is difficult. Therefore emphasis has to be made keeping in mind the demand for goods and other items. A long term perspective in selected area of production will enable the profitability and sustainability of agriculture including pest management.

References

Jayaraj, S., A.V. Rangarajan, R.J. Rabindra and N. Chandra Mohan (1991). Studies on the ecology of certain soil-borne insect pests in Tamil nadu State, India. In: Advances in Management and conservation of soil fauna ed. G.K.Veeresh *et.al.*, New Delhi, Oxford& IBH Publishing Co. Pvt.Ltd., pp.183-190.

Murthy, R.L.N. and P. Venkateswarulu (1998). Introducing eco- friendly farming techniques and inputs in cotton. In proceeding of the workshop on "Eco-friendly Cotton, 1998", Indian Society for Cotton Improvement, Mumbai, Oct.31,1998 pp.29-32.

11

Nematode Problem in Pulses and Their Management

Virendra Kumar Singh

Nematodes are the most numerous metazoa on earth. They are either free living or parasites of plants and animals. Although currently only about 4100 species of plant parasitic nematodes have been described (i.e. 15 % of the total number of nematode species known), their impact on humans by inflicting heavy losses in agriculture is substantial. Most of the developing countries including India lie in tropical or sub tropical regions where climate is suitable for activity and multiplication of nematodes almost throughout the year. Sandy and warm soils such as found in India other developing countries in the arid zone are very favorable for nematode infection, especially in irrigated areas which are used for crop production continuously. The percentage of crop losses caused by plant pathogens, insect pests and weeds world wide has been estimated to 42% according for $500 billion worth of damage. The estimated overall annual yield loss of world's major crops due to damage by plant parasitic nematodes has been reported to the extent of 12.3% (Sasser & Freckman, 1987). Estimated losses due to nematodes attack on different cultivated crops all over the world by FAO are around 400 million dollars. Annual crop losses in India is Rs. 242.1 billion. A large number of plant parasitic nematodes have been found associated with pulse crops in allover India but pathogenicity has not been proved for all. Plant parasitic nematodes are one of the most unwanted organisms in agricultural soils. They parasitize the roots of our carefully nurtured crops, reduce their ability to produce yields, make them weak and perhaps vulnerable to many pests and diseases. They reduce the crop yields that may equal those associated with more severe and easily identifiable plant diseases and we do not even realize their presence. Damage to crops by nematodes often goes unnoticed or attributed to other causes such as lack of fertility, deficient soil moisture. Sandy and warm soils such as found in India other developing countries in the arid zone are very favourable for nematode infection, especially in irrigated areas which are used for crop production continuously (Taylor, 1967).The detail work have been done under pulses improvement program at ICRISAT by Sharma (1985).Time to time several workers have reported various type of nematodes on pulses which cause severe damage to crops. These nematodes have associated with pulses such as *Heterodera* spp., *Meloidogyne* spp., *Rotylenchulus* spp., *Hoplolaimus* spp., *Pratylenchus* spp., *Helicotylenchus* spp., *Tylenchorhynchus* spp. etc.

Root-Knot nematodes: (*Meloidogyne* spp.)

Root-knot nematodes are important pests of pulse crops such as Pea, cowpea, bean, black gram, green gram, chickpea, lentil, pigeonpea and several other leguminous crops. India about 12 species have been reported to occur. But, of all these species, predominating populations encountered are of *M. incognita*, *M. javanica*, *M. arenaria* and *M. hapla*. Bio-chemical differences in protein and enzyme content among species have also been used for taxonomic identification. These four species account for more than 95% of the root-knot nematode population world over (Sasser & Carter, 1985). The above ground symptoms: Root knot nematode damage results in poor growth, a decline in quality and yield of the crop and reduced resistance to other stresses (e.g. drought and other diseases). A high level of root-knot nematode damage can lead to total crop loss.

Nematode damaged roots do not utilize water and fertilizers as effectively leading to additional losses for the grower. Below ground symptoms: The most distinctive nematode symptoms on roots are galls caused by *Meloidogyne* spp. They are small, individual, bead like or fusiform swellings in some roots. In other plants, galls may be massive lumps of fleshy tissue more than 1 inch in diameter containing dozens of nematodes. Other root swellings must not be mistaken for root-knot galls. Nitrogen fixing bacteria cause swellings on the roots of most legumes. These swellings called nodules are easily distinguished from root-knot galls by differences in how they are attached to the root and their contents. Nitrogen nodules are loosely attached to the root, and can generally be very easily removed, root knot galls originated from infection at the center of root, so they are an integral part of the root whose removal requires tearing the cortex apart. In addition, fresh nodules should have a milky pink to brown liquid inside them, while root-knot galls have firmer tissues and contain female root-knot nematodes inside the gall tissues, near the fibrous vascular tissues of the root. The male and female root-knot nematodes are easily distinguishable morphologically. The males are wormlike and about 1.2 to 1.5 mm long by 30 to 36 μm in diameter. The females are pear shaped and about 0.40 to 1.30 mm long by 0.27 to 0.75 mm wide. The life cycle includes egg, juvenile and adult stages. A life cycle is completed in 25 days at 27°C, but it takes longer at lower or higher temperatures. The female lays about 400-500 eggs per in a gelatinous matrix secreted by rectal glands. The root knot nematode, *M.incognita* on pea and bean was reported first time from farmer's field of Jammu (Singh and Satpriya, 2008) and also *M.incognita* on Lentil (Singh et al., 2008). A mixed infestation of *M.incognita* and *M. javanica* reduced seed yield 39% (Honda and Mishra,1989). Various types of work viz., penetration, development, biology and pathogenicity have studied of *M.incognita* on pea (Singh, 2009 and Singh, 2010)

Cyst nematodes: (*Heterodera* spp.)

The nematodes belonging to the genus Heterodera are called cyst nematodes because at the time of death the mature females get converted into a brown cyst in which the eggs are held. Various species of *Heterodera* are popularly known as the cyst nematodes. The above ground symptoms resemble these associated with root damage and include stunting of shoot, yellowing of foliage and reduced size of various shoot parts. An experienced observer can often see cyst nematode, *Heterodera* spp. on the roots of their hosts without magnification. The young adult females are visible as tiny white colour. After a female cyst nematode dies, her white body wall is tanned to a tough brown capsule containing several hundred eggs. The severity depends upon density of nematodes attacking roots. The mature female bodies are found attached to roots by their head end embedded almost in the stele. The site of feeding is modified into a syncitium similar to that found in case of *Meloidogyne*. On the surface of infected roots, white to brown bodies of females can be discerned with naked eyes. The intensity of body colour depends upon the maturity stage of the young female or cyst. In case of cyst nematode matrix with eggs may also be found attached to the posterior region of the female body. The male is wormlike, about 1.3 mm long by 30 to 40 m in diameter. Fully developed females are lemon shaped, 0.6 to 0.8 mm in length and 0.3 to 0.5 mm in diameter. Approximately 21-30 days is required for the completion of life cycle of this nematode. Several workers have studies the effect of initial inoculum level of *Heterodera* spp. on pulse crops (Sharma and Sethi, 1976, Dalal and Bhatti, 1989). Several types of work such as penetration, development, volume, biomass, biology and pathogenicity have been studied of *Heterodera cajani* on pigeonpea (Singh and Singh 1988, Singh 1992, Singh and Singh 1997, Singh and Singh 1998, Singh and Singh 1999).

Lesion nematodes: (*Pratylenchus* spp.)

Root Lesion nematodes occur in all parts of the world. Assessment of exact loss by lesion nematodes has not been made possible under field conditions due to presence of mixed population of the nematode in the field. Lesion nematodes have wide host range which can affect the selections of crop used to control the nematode in crop rotation sequences. Soil type tillage operations have also been reported to affect lesion nematodes population dynamics. The plants show chlorosis, stunting and general lack of vigour resulting into wilt. The plant form patches or zones in the field. The roots show necrosis and lesions which become ideal for infection of other microorganisms. The presence of small brown to black lesions on the root surface is the most important symptoms or damage produced by the lesion nematode. Both male and female of these nematodes are wormlike, 0.4 to 0.7 mm long and 20 to 25 m in diameter. They are migratory endo parasitic nematodes. The life cycle of the various species of *Pratylenchus* is completed within 45 to 65 days. Temperature, soil type, moisture and tillage operations are important environment factors which greatly affect the development and reproduction of nematode species as well as disease development.

Reniform nematodes: (*Rotylenchulus reniformis*)

Reniform nematode is pest of a large number of pulses in India, it is considered as pest of great significance after root-knot nematode. The nematode is a semi-endoparasite and remains attached to roots. Infected plants show stunting in growth with reduced and discolored root system. Damage during pre and post emergence of seedlings leads to reduction in germination and crop stand. The total duration of life cycle is about 4-5 weeks under optimum conditions. Soil P^H is also an important factor affecting reproduction of the nematode. Soil moisture and temperature have profound influence on infectivity, penetration and the biology of the nematode. There was significant reduction in plant height, shoot and root weight of mash at 1000 or more juveniles/ 500 g soil.

Lance nematodes: (*Hoplolaimus* spp.)

Lance nematodes are ectoparasites. Some times they feed at a particular site for a long time with nearly half of the body inside the rot system (sedentary ectoparasite). In many cases, juveniles of the lance nematode completely enter the cortical tissue (endoparasite). *Hoplolaimus* spp. multiplies slowly in comparison to endoparasitic nematodes but they inflict significant crop damage at a lower level of infection. Several workers have reported significant reduction by the *Hoplolaimus* spp. on various crops in the different parts of India (Haider *et al.* 1978, Gaur and Mishra, 1981; Das, 1982; Mulk and Jairajpuri, 1975; Sharma, 1985). The availability of feeder roots and temperature are important factors for population build up of this nematode (Haider and Nath, 1992).

Cultural Management

The cultural practices include crop rotation, fallowing, flooding, sanitation, ploughing during summer season, mulching, organic manure, spacing of plants in the field, time of sowing, resistant varieties etc.One of the most effective eco-friendly management practices is the use of proper combination of cultural practices. The affectivity of cultural method depends upon the proper knowledge of life history, population dynamics host range of the plant parasitic nematodes infesting crops. Nematode management through inter cropping and crop rotation is based on the fact that some species of nematodes are able to feed and multiply only on host crop.Rotation is a very old practices for reducing soil borne problems. Rotation to non-host crops may cause many of those pests to cease reproduction and allow natural mortality factors to reduce their numbers. Flooding or following may be used to help reduce numbers of nematode pests. The period of flooding appears to vary with several factors such as kind of soil, season etc. Juveniles (larvae) are more easily killed by flooding than eggs. The period of flooding needs to be worked out for each condition. Fallow periods in cropping sequences can also reduce nematode populations. The summer ploughing 2-3 times during hottest period of the year help to expose nematodes to the drying action of sun and wind and reduce the population. Haque and Prasad (1982) reported drastic reduction of plant parasitic nematodes and significantly increased the yield with three deep ploughing. Singh and Singh (1999) studied the different inoculum levels of *H. cajani* on pigeonpea. They found that various growth parameters were decreased in different inoculum levels such as 2000, 1000, 500 and 100 juveniles/pot in comparison to without inoculated pot. The intensity of symptoms was directly related to the inoculum levels. In general, the entry of the nematode in roots of pigeonpea was higher at higher levels of inoculum. The nematode multiplication was universally proportional to inoculum density size of the egg sac was more at lower inoculum with higher number of eggs/egg sac. Singh and Singh (1995) investigated development and pathogenicity of *Heterodera cajani* on susceptible and moderately susceptible cultivars of pigeonpea. The penetration by second stage juveniles in roots of ICPL-87, susceptible cultivar was earlier i.e. within 24 hours as compared to moderately susceptible cultivar PDA –I where penetration occurred only after 24 hours. The different stages of *H. cajani* were found earlier in susceptible cultivar than the moderately susceptible one. All the stages of the nematode persisted longer in cv. PDA-I than ICPL-87. Number of eggs /egg sac was also lower in the moderately susceptible cultivar as compared to susceptible cultivar of pigeonpea. Reduction in plant growth characters was more in susceptible cultivar ICPL-87 than the moderately susceptible cultivar PDA-I to nematode infection. Multiplication of *H. cajani* was higher in ICPL-87 than in PDA-I as higher number of cysts, egg sacs, eggs and juveniles/pot.

Sanitation terms covers a wide range of cultural practices, including weed control, crop residue destruction and disinfestations of farm equipment before moving it from heavily infested fields to uninfected fields. In monocultures, eliminating the weed hosts can be important in reducing the populations of plant parasitic nematodes. Soil temperature plays crucial role in the activities of plant parasitic nematodes, the time during which crop is planted is important (Ayaub, 1980). Singh and Singh (1995) studied the effect of time of inoculum on growth of pigeonpea and biology of *Heterodera cajani*. The pathogenic effect of the nematode on the plant decreased with increasing time of inoculation after germination of the seed. The pathogenic effect was minimum in 28th day old plant inoculated with nematode. The multiplication of the nematode was maximum in 7 day old plant inoculated followed by 14 day, 21 day. Minimum multiplication was recorded in 28th day plant and than o day. The development of the nematode was faster in 7 day old inoculated plants while the persistence of their different stages of the nematode was minimum. Most pathogenic nematodes are inactive at lower temperature. Manipulation of date of sowing of chickpea escaped root-knot nematode, lesion nematode and raniforam nematode was reported by (Gaur et al. 1979, Anon, 1999).

The addition of inorganic fertilizers alone without organic manure usually increases the nematode population and disease intensity. NH_4-N reduces the disease incidence while NO_3-N may increase the same. Particular forms as well as dose and proportion of NPK may also reduce or increase the incidence of disease. Use of tolerant/resistant varieties is most practical approach for the management of nematode diseases. Crop cultivars resistant to phytonematodes can be the most useful and cheapest means of nematode control for the small-scale farmers.

Nematicidal plants with roots containing nematicidal substances have been investigated. These toxic substances reduce the population level of some nematode species. African marigolds (*Tagetes* spp.), asparagus, crotalaria, mustard and several cruciferous plants have been reported to produce toxic substances. Soil amendments with green manure, compost, oil cakes of neem, mahua, mustard, groundnut, cotton, linseed, karanj and saw dust etc. have been found to reduce nematode populations. Neem, karanj and groundnut cakes incorporated into soil at the rate of 1,500 kg/ha give good control of plant parasitic nematodes and could be practiced wherever possible. Apart from encouraging the multiplication of natural enemies like nematode trapping fungi, the decomposition products of these organic amendments are toxic to nematodes. Use of organic manures is of great value since the decomposition products and promotion of natural enemies decrease nematode populations (Singh and Sitaramaiah, 1973; Singh and Singh, 1991; Singh and Singh,1992; Alam, 1990;

Haseeb et al., 1984a; Mankau, 1963; Mishra and Prasad, 1974; Singh and Singh, 2001; Singh, 2004; Singh, 2008; Ononuja and Kpodobi, 2008) reported that among the three cakes used i.e. neem cake, linseed cake and mustard cake each @ 5q/ha neem cake reduced the population *viz.*, cysts, males, egg-sacs, eggs and Juveniles per pot of *Heterodera cajani* on pigeonpea, more as compared to others. Haque *et al* (1997) reported significant reduction in infections of *M. incognita* on mungbean, when soil amended with *Calotropis procera* leaves @ 0.5 and 1.0 percent (w/w). Patel and Patel (1992) reported significant increase in plant growth with 73% reduction in *Rotylenchlus reniformis* population infecting pigeonpea, when treated with poultry manure @ 0.720g/pot with 800g soil followed by press mud @ 1.198g/pot. Singh (2008) reported significant reduction infection of *M. incognita* on pea, when soil amended with various organic amendments @ 2.5 and 5.0 (w/w). The amount of oil cakes or any organic matter to be incorporated depends on various factors like, soil type and texture, crop to be planted, the predominant nematode fauna of the soil and the amount of soil moisture present in the soil. In trap crops when grown in infested soils, the nematodes penetrate into the root system and start multiplying. Before the nematodes complete its life cycle the plants are uprooted and destroyed. Applications of botanicals are easy, environmentally safe, and having no phytotoxic effect on crops. The inhibition of root-knot development may be due to the accumulation of toxic by products of decomposition (Alam *et al,* 1979) to increased phenolic contents resulting in host resistance (Alam *et al* 1979, 1980), or to changed physical and chemical properties of soil inimical to the nematodes (Ahmad *et al*, 1972). One of the most economical and effective ways to control plant parasitic nematodes is the growing of nematode resistant plant cultivars. The identification of a dominant Mi gene importing resistance to the root-knot nematodes and its linkage to an acid phosphates gene is a major break through in this area of research. Several varieties of pulse crops were listed to see their resistance against plant parasitic nematodes.

Biological Management

Application of natural enemies of plant parasitic nematodes for controlling nematode population is an essential component of eco-friendly management. Nematologists in all over the world are working very hard to identify and lean to manipulate natural enemies of nematodes so they can be used as biological control agents. Nematodes have many natural enemies including fungi, bacteria and predacious nematodes. Certain fungi capture and kill nematodes in the soil. *Arthrobotrys* spp., *Dactylaria* spp., *Dactylella* spp., *Catenaria* spp., and *Trichothecium* spp., are the genera most commonly represented. Siddiqui and Mahmood (1996) reported significant reduction infection of the *H. cajani* and *Fusarium udum* on pigeonpea, when applied bioagents. Kumar and Prabhu

(2008) reported significant increase in yield with 52-73% reduction in cyst population infecting pigeonpea. When used *Trichoderma harzianum* and *Pochonia chlamydosporia*. The biological control of root knot nematode on tomato under green house condition by using predaceous fungi has been reported Singh *et al* (2001). Some fungi capture nematode by adhesion, but many employ specialized devices that include networks of adhesive branches, stalked adhesive knobs, non constricting rings and constricting rings. The surface of the nematode is penetrated and the fungus hyphae grow throughout the nematode body, digesting and absorbing its contents. Under favourable conditions, large numbers of nematodes may be captured, and killed especially by those fungi that form adhesive net works or hyphal loops. Wide range of opportunities fungi parasitizes eggs and cysts of nematodes. *Trichoderma harzianum, Trichoderma virens, Aspergillus niger, Paecilomytes lilacinus, Verticillium chlamydosporium* are found promising biocontrol agents. Now mycorrhiza is not restricted to its use only as biofertilizers, its potential role in the biological control of plant parasitic nematodes is reported by many workers. Sikora (1979) found that prior presence of VAM fungi *Glomus mosseae* has resulted into an increase in plant resistance against *Meloidogyne* spp.

A bacterial parasite of nematodes *Pasteuria penetrans*, has received much attention and research effort in recent years, *P. penetrans* is probably the most specific obligate parasite of nematodes, with a life cycle remarkably well adapted to parasitism of certain phytonematodes. It directly parasitizes juvenile nematode, thus affects penetration and reproduction. Seed bacterization, soil drenching and bare root dip application with *Pseudomonas fluroscenes*, *Bacillus subtilis*, *B. polymyxa* effectively controls plant parasitic nematodes was also reported by many workers. Gogoi and Neog (2003) reported *pasteuria* treated plots significantly root-knot nematode population on green gram, cowpea and french bean. Among the predatory nematodes, monarchs may be proved efficient predators because of stronger predatory potential, high rate of predacity and high strike rate (Bilgrami, 1998).

Chemical Management

Nematicides are chemicals used to control plant parasitic nematodes. Two major groups of nematicides are distinguished by the manner in which they spread through the soil. Soil fumigants are gases in cylinders or liquids which spread as gases from the point where they are infected into the soil. Non fumigant nematicides include a variety of water soluble compounds which are applied to the soil as liquid or granular formulations. Most belong to the carbamate or organophosphate families of pesticides. These distributions in the soil depend on physical mixing during application and moving in solution in soil water. There is no perfect nematicide for all purposes. Nematicides vary in their effectiveness against different kinds of nematodes, ease to handling, cost effect on other classes of pests (weeds, disease organisms, insects) behavior in different soils, toxicity to different plants and availability. The performance of the nematicides will depend on soil conditions, temperatures and rainfall. A yield benefit is not guaranteed and nematicides are expensive.

The older nematicides are mostly fumigants which are applied in soil. These are seldom recommended because of their hazardous nature and high toxicity to non target organisms. Recently, large number of non fumigant and systemic nematicides are available which are safe on plant. DBCP (nemagon) has been found very effective for standing crops against root-knot nematodes but its use has been suspended due to adverse effect on human beings. Carbofuran 3G (furadon), phorate 10G' (thimet), fenamiphos (nemacur), fensulphothion (dasanit) have been recommended for a variety of crops. Application of some granular nematicides such as phorate, aldicarb and carbofuran @2.5 mg a.i/kg soil (5.0 kg a.i/ha.) reduced the entry of *H. cajani* in roots of pigeonpea. The development of the nematode was also delayed by all these nematicides. Maximum adverse effect on penetration and development of *Heterodera cajani* was produced by aldicarb followed by carbofuran and phorate. All the nematicides delayed egg laying and reduced the number of eggs/egg sac was found in aldicarb. All nematicides decreased the final population of nematode. The browning of cysts was also delayed by all nematicides (Singh and Singh 1998).Singh (2008) reported significant reduction in root - knot nematode, *M .incognita* population of pigeonpea,when soil amended with carbofuran and phorate nematicides @1.0 and 1.5kg /ha. However, growth of the pigeonpea was maximum in both nematicides at both doses of each as in comparing to control. The dose and method of application would vary with crop. Chemicals and Phytoalexins that have implications in the nematode control (Prasad and Swarup 1986), synthetic pheromones to disrupt nematode reproduction Bone (1987), and chemicals that directly interfere with nematode chemoreceptors Janson (1987) should be exploited for nematode population management. Prasad (1990) studied the control by seed pelleted with carbofuran, aldicarb, sulfone @ 3 and 6% w/w. Application of carbosulfan nematicides reduced root-knot nematodes in pulses and seed treatment was more effective than soil application.

Regulatory Management

Numerous attempts have been made to prevent the introduction of nematodes into countries or provinces by means of quarantine. Quarantines are established by legislative action in parliament, etc., and usually give quarantine authorities power to make and enforce regulations to accomplish the purpose. Such regulations usually prohibit bringing infected seeds into protected areas where similar crops might become infected.

Physical Management

Physical means of nematode control includes heat treatment of soil, solar drying, steam sterilization, hot water treatment and soil solarization. Soil solarization with transparent polyethylene sheet has been attended as a means of raising the soil temperature to lethal levels to control soil pathogens (Sharma and Nene, 1990) Soil solarization, a method of pasteurization can effectively suppress most species of nematode along with other microbes and weeds under field conditions (Rao and Krishappa, 1995, Kumar *et al.* 1993). Generally, sheets of 50-100 μm thickness are most suitable for raising the soil temperature. The use of transparent polyethylene had yielded better results than black sheets, since transparent sheet transmit most of the incident radiation to soil (Mehrer, 1979, Mehrer and Katan, 1980). Mutagenic treatment with EMS and MH etc. were found effective for inducing resistance in black gram (Routaray *et al.*, 2001). Additions of salt, sugar, charcoal etc. create osmotic stress on nematodes and can be used for controlling nematodes.

Integrated Management

The integrated nematode management is based upon the system approach, follows location specific principle and is environment specific (Pankaj and Sharma, 1998). Utilization of the best combination of available management strategies for the pest complex at hand (nematodes, insect pests, disease organisms, weeds etc.) constitutes an integrated crop protection system. Resistant cultivars, crop rotation, pesticides, and sanitary and cultural practices can all be employed to the best possible advantage. An integrated management strategy prevents the excessive build up any single nematode, insect, or disease population and minimizes the development of pest resistance to any single tactic. Integrated pest management systems require flexibility and depend upon the specific pest problem and locally available management options. A fixed set of recommendations may keep a pest complex in check for a limited period of time, but as the pest population shifts, recommendations will have to change also. Therefore, system development takes into account many factors including the species and race of pests present, the availability of resistant host plants, the longevity of the pest, and the crops, cropping systems, and climate of the geographical region. The end result is a management strategy tailored to fit the unique circumstances of each pest situation.

Application of Molecular Biology and Biotechnology

Advances in molecular biology and biotechnology provide means to develop novel strategies for nematode management. Nematode diagnostics is essential for success of any nematode management programme even more when the control method is highly specific to the control species/pathotypes/race. A number of new techniques for analyzing nucleic acids, proteins, carbohydrates and lipids can be helpful in the identification of pests of those allozyme; monoclonal antibody and DNA based systems are most well developed for nematodes. Polyacrylamide gel electrophoresis (PAGE) has mainly focused on cyst and root-knot nematodes (Davies, 1977). PCR techniques or RFLP analysis used as supplementary tools wherever necessary. Nematode resistant transgenic plants can be designed by various approaches. The simple method is to introduce resistant genes effective against plant parasitic nematodes from wild species to commercial cultivars.

Conclusions

Plant nematodes parasitize the roots of our carefully nurtured crops, reduce their ability to produce yield, make them weak and perhaps vulnerable to many pests and diseases. There is a need to reduce these avoidable yield losses by developing new environment friendly, economically acceptable and ecologically based management strategies as the current options become ineffective or unacceptable. Crop resistance offers to be of great importance in this directivity. In addition, an intensive search of plant germplasm collection for natural resistance to nematodes and their races is also required. An in depth understanding is needed of the molecular basis of how and why plants are susceptible to nematodes. New progress is being made in studying changes in gene expression during the infection of plants in nematode host interactions where feeding sites are formed. The research on identifying the promising biocontrol agents, their mass production, application techniques and their behavior in the soil under varying agro climatic conditions need to be intensified.

References

Ahmad, R., Khan, A.M and Saxena, S.K. (1972). Changes resulting from amending the soil with oil cakes and analysis of oil cakes (Abstract) Proc. 59th Ses. Indian Sci. Con.Calcutta. Part III, PP. - 164.

Alam, M.M., Kahn, A.M., and Saxena, S.K. (1979). Mechanism of control of plant parasitic nematodes as a result of the application of organic amendments to the soil. V. role of phenolic compounds. *Indian J. Nematol.* 9: 136-142.

Alam, M.M. (1990). Control of plant parasitic nematodes with organic amendments and nematicides in nurseries of annual plants. *J.Bangladesh Acad. Sci.* 14:107-113.

Anonymous (1999). All India Coordinated Research Project on "Plant parasitic nematodes with integrated approach for their control, ICAR Biennial report (1997-99), C.C.S. Haryana Agricultural University, Hissar.

Ayaub, S.M. (1980). Plant Nematology: An Agricultural Training and Nema aid publication Sacramento, C.A. P-195.

Bandyopadhyay, P. Kumar, D. Singh, V.K. and Singh, K.P. (2001). Eco-friendly management of root-knot nematode of tomato by *Arthrobotrys oligospora* and *Dactylaria brochopaga*. *Indian J. Nematol.* 31 (2): 153-156.

Bilgrami, A.L. (1981). Predatory nematodes and protozoans as biopesticides of plant parasitic nematodes. In plant nematode management a biological approach, (Ed.) by P.C. Trivedi. PP 4-23.

Bone, I.W. (1987) Pheromone communications in nematodes. Pages 147 – 152 In vistas on Nematology. A commemoration of the twenty fifth anniversary of the Society of Nematologists. (Veech, J.A. and Dickson, D.W., eds.), Hyattsville, Maryland, U.S.A. Society of Nematologists.

Dalal, M.R. and Bhati, D.S. (1989). Pathogenicity of *Heterodera cajani* in mungbean and cluster bean as affected by presence or absence of Rhizobium. *Indian J. Nematol.* 19: 153-158.

Gaur, H.S., Mishra, S.D. and. Sud, U.C (1979). Effect of date of sowing on the rotation between the population density of root-knot nematode, *M. incognita* and the growth of three varieties of chickpea. *Indian J. Nematol.* 9: 152-159.

Gaur, H.S., Mishra, S.D. (1981). Pathogenicity of the lance nematode, *Hoplolaimus indicus* to cotton. Indian J. Nematol. (1): 87-88.

Gogoi, B.B and Neog, P.P. (2005). Efficacy of *Pasteuria penetrans* against *Meloidogyne incognita* on green gram under field conditions. *Indian J. Nematol.* 35: 80-81.

Haider, M.G. and Nath, R.P. (1992). Curr. Nematol. 3: 173-177.

Haider, M.G., Nath, R.P., Prasad S.S. (1978). Studies on the lance nematode *Hoplolaimus indicus I-* pathogenicity and histopathogenesis on maize. *Indian J. Nematol.* 8 (1): 9-12.

Haseeb, A., Alam, M.M. and Khan, A.M (1984a) Control of plant parasitic nematodes with chopped plant leaves. *Indian J. Plant Pathol.* 2: 180-181.

Haque, M.M. and Prasad, (1982). Effect of agronomical practices and DD application to nematodes and crop yield. Indian J. Nematol. 13: 126-129.

Haque, S.E., Sultana, V., Abid, M., Ara, J. and Ghaffer, A. (1997) Use of *Calotropis procera* and microbial antagonists in the control of *Meloidogyne* root knot on Mungbean Pak. *J. Phytopathol.* 9: 108-110.

Jansson, H.B. (1987). Receptors and recognition in nematodes. Pages 153-158 in vistas on Nematology. A commemoration of the twenty fifth anniversary of the Society of Nematologists. (Veech, J.A. and Dickson, D.W., eds.), Hyattsville, Maryland, U.S.A. Society of Nematologists.

Kumar, B.N., Yaduraju, T., Ahuja, K.N. and Prasad, D.P. (1993). Effect of soil solarization on weeds and nematodes under tropical Indian conditions. *Weed Research*, 33: 423-429.

Kumar, S. and Prabhu, S. (2008). Biological control of *Heterodera cajani* in pigeonpea by *Trichoderma harziaum* and *Pochonia chlamydosporia. Indian J.Nematol.*.38 (1): 65-67.

Katan, J. (1981). Solar heating (solarization) of soil for control of soil borne pests. *Ann.Rev. Phytopathol.* 19:211-236.

Mehrer, Y. (1979). Prediction of soil temperature of a soil mulched with transparent polyethylene. *J. Appl. Meteorol.* 18: 1263-1267.

Mehrer, Y. and Katan, J. (1980). Prediction of soil temperature under polythene mulch Hass. 60: 1384-1387.

Mankau, R. (1963) Effect of organic soil amendments on nematode population. *Phytopathology*. 53: 881-882.

Mishra, S.D and Prasad, S. K (1974). Effect of soil amendments on nematodes and crop yields. *Indian J. Nematol.* 4:1-19.

Mulk, M.M. and Jairajpuri, M.S. (1975). Nematodes of leguminous crops in India III three new species of *Hoplolaimus. Indian J. Nematol.* 5 (1): 1-8.

Ononuju, C.C and Kpadobi, U.C. (2008). Effect of plant leaves on the control of *Meloidogyne incognita* in soybean. *Indian J. Nematol.* 38 (1): 1-4.

Pankaj, and Sharma, H.K. (1998). IPM strategies for nematode management. In potential IPM tactics (eds. D. Prasad and R.D. Gautam) Westvill Publishing House, Paschim vihar, New Delhi, PP. 131-145.

Patel, R.G. and Patel, D.J. (1992). Comparative efficacy of some organic amendments and granular nematicides in the control of *Rotylenchulus reniformis* infecting pigeonpea. *Indian J. Nematol.* 22: 14-18.

Prasad, D. (1990). Control of root-knot nematode *Meloidogyne arenaria* on groundnut by chemical seed pelleting. *Curr. Nematol.* 1:31-34.

Prasad, D. and Yaduraju, N.T, (1993). Solarization of filed soil for effective management of nematodes and weeds. National conference on ecofriendly approaches, Dec. 20-22, 67, (Abstr.).

Rao, M.S. and Reddy, P.P (1992). Prospects of management of root-knot nematode on tomato through the integration of bio control agent and botanicals. Seminar on current trends in diagnosis and management of plant diseases. IIHR. Bangalore, PP-10.

Rao, V.K. and Krishappa, K. (1995). Soil solarization for the control of soil borne pathogen complexes with special reference to *Meloidogyne incognita* and *Fusarium oxysporium*. Indian Phytopathol. 48: 300-303.

Routaray, B.N., Dash, S.B. and Mishra, R.C. (2001). Effectiveness of mutagenic treatment in induction of resistance to *M. incognita* infection in black gram. National Congress on Centenary of Nematology in India: Appraisal and future plans .5-7 Dec, 2001, I.A.R.I., New Delhi, PP. 154-155 (Abstract).

Sasser, J.N. and Carter, C .C. (1985). An advanced treatise on *Meloidogyne*. Vol I: Biology and Control. North Carolina State University, Graphics pp.1-422.

Sasser, J.N. &Freckman, D.W. (1987). A World perspective on Nematology. In: vistas on Nematology (eds. Veech, J.A. and Diskson, D.W.) Pub. by Society of Nematologist, U.S.A pp. 7-14.

Sharma, S.B. (1985A). Nematode diseases of chickpea and pigeonpea. Pulse pathology progress report No. 43 Patancheru, A.P. (India) International Crops Research Institute for the Semi Arid Tropics pp. 103.

Sharma, S.B. and Nene, Y.L. (1985). Effect of presowing solarization on plant parasitic nematodes in chickpea and pigeonpea fields. *Indian J. Nematol.* 15: 277-278.

Sharma, S.B. and Nene, Y.L. (1990). Effects of soil solarization on nematodes parasitic to chickpea and pigeonpea. *J. Nematol.* 22: 658-664.

Sharma, V.K. and Sethi, C.L. (1976). Interrelationship between *Meloidogyne incognita*, *Heterodera cajani* and *Rhizobium* spp. on cowpea. *Indian J. Nematol.* 6 (2): 117-123.

Sikora, R.A. (1979). Predisposition to *Meloidogyne* infection by the endo tropic mycorrhizal fungus *Glomus mosseae*. In root-knot nematode Meloidogyne spp. systematic biology and control (eds.F.Lamberti and C .Taylor Academic Press London) PP.399-404.

Singh, R.S. and K. Sitaramaiah (1973) Res. Bull. No. 6, G.B. Pant University of Agril. Sci. & Tech., pp 1-289.

Singh, K.P. and Singh, V.K. (1988). Effect of time of inoculation with standard inoculum of *Heterodera cajani* on the manifestation of symptoms in pigeonpea. Indian Phytopathol. 41: 279 (Abstract).

Singh, K.P. and Singh, V.K. (1991). Nematicidal natures of Arjun bark *Terminelia arjuna* on cyst nematode, *Heterodera cajani* of pigeon pea. *New Agric.* 2(1): 77-78.

Singh, V.K. (1992). Biology of *Heterodera cajani* associated with pigeonpea roots with emphasis on pathogenesis. PhD thesis, B. H.U.,Varanasi (India).

Singh, K.P. and Singh, V.K. (1992). *Terminalia arjuna* leaf powder reduces population density of *Heterodera cajani. International Pigeonpea Newsletter* 16:17-18.

Singh, V.K. and Singh, K.P. (1995). Effect of time of inoculation on growth of pigeonpea and biology of *Heterodera cajani. Curr. Nematol.* 6 (2): 107-115.

Singh, K.P. and Singh, V.K. (1997). Studies on volume and biomass of *Heterodera cajani* on susceptible and moderately susceptible cultivars of pigeonpea. *Int. Tropical Plant Diseases*. 13: 237-244.

Singh, V.K. and Singh, K.P. (1998). Effect of granular nematicides on the biology of *Heterodera cajani* of pigeonpea. *Indian J. Nematol.* 28 (2).168-173.

Singh, V.K. and Singh, K.P (1998). Penetration and development of *Heterodera cajani* on pigeonpea with reference to its biology. *Acta Botanica Indica* 26: 47-53.

Singh, V.K. and Singh, K.P. (1999). Effect of different inoculum levels of *Heterodera cajani* on biology of nematode and plant growth of pigeonpea. *Curr. Nematol.* 10(1):1-12.

Singh, V.K. and Singh, K.P. (2001). Effect of some medicinal plant leaves on the biology of *Heterodera cajani*. *Indian J. Nematol*. 31(2):143-147.

Singh, V.K., (2004). Management of *Heterodera cajani* on pigeonpea with nematicides and organic amendments. *Indian J. Nematol*. 34: 213-214.

Singh, V.K. (2008) Effect of organic amendments on the biology and management of root-knot nematode (*M. incognita*) in pea. IPS-MEZ Annual meeting & National Symposium on advances in microbial diversity and disease management for sustainable crop production. 13-15, Oct., 2008, (Abstract).

Singh, V.K. (2008). Management of root-knot nematode, *Meloidogyne incognita* infecting pigeonpea. *Indian J. Nematol.* 38 (1):112-113.

Singh, V.K. (2008). Eco-friendly management of plant parasitic nematodes in vegetable crops. P.440-447 Insect pest and disease management (Ed. Dr. D. Prasad) Daya Publishing House, New Delhi.

Singh, V.K (2008). Effect of soil solarization for management of plant parasitic nematodes. *Ann. Plant Prot. Sci.* 16 (2) 541-542.

Singh, V.K. V. Koul, C.S. Kalha, V.B. Singh, and Verma, V.S. (2008). New record of root-knot nematode *Meloidogyne incognita* infecting lentil in Jammu. *Journal of Research* 7: (2) 267-268.

Singh V.K. (2009). Pathogenic effect and management with organic amendments of root-knot nematode, *Meloidogyne incognita* infecting pea. 5th International Conference on Plant Pathology in the globalized era 659 (S15) P.353 (Abstract).

Singh V.K (2010). Penetration and development of *Meloidogyne incognita* on pea with reference to its biology. *Journal of Research*, 9(2): 250-254.

Taylor, A. (1967). Introduction to research on plant nematology. FAO guide to the study and control of plant parasitic nematodes, pp. 133.

Upadhyay, K.D., Dwivedi, K. and Azad, C.S. (1987). Analysis of crop losses in pea and gram due to *Meloidogyne incognita. International Nematology Network Newsletter,* 4 (4): 6-7.

12

Approaches in Pest Management of Stored Grain Pests

Ankit Kumar, Surender Singh Yadav and Manoj Kumar Jat

A large number of insect pests are associated with stored grains which are directly related to geographical and climatic conditions. There are different estimates on post harvest losses in food. During storage both quantitative and qualitative losses occur due to insects, rodents, and microorganisms. Almost all the insect species may destroy 10.0 - 15.0 % of grain and contaminate with undesirable odour. These pests also helps in transportation of fungi. Some insect pests initiate damage at the ripening stage of crops and continue during storage. Major sources of infestations are old gunny bags, storage structure and old containers (Pruthi and Singh, 1950). The spread and distribution of stored product pests are facilitated by movement of grains from one area to another and by flight of insect pests as some of the adult insects are strong fliers. The cracks and crevices are probable sites for cross infestation. Nearly one thousand species of insects have been associated with stored products in different part of the world. Majority of insect pests belong to the orders Coleoptera and Lepidoptera (Khare, 1994).

There are two types of pests which damage the grains in store

Primary insect pests: Insects which cause damage to stored grains by directly feeding on the grain at some point in their lifecycle. Primary pest species often develop and reproduce very quickly when the conditions are optimal and the large populations cause considerable damage within few months. Many beetles feed internally in grain kernels as larvae. Rusty grain beetles, weevils, and lesser grain borer all develop initially inside the kernel. The primary stored grain insect pests include lesser grain borer, larger grain borer, pea weevil, Southern cowpea weevil, granary weevil, bean weevil, rice weevil, maize weevil, rusty grain beetle, flat grain beetle, flour mill beetle, merchant grain beetle, sawtoothed grain beetle, khapra beetle, longheaded flour beetle, confused flour beetle, red flour beetle, large flour beetle, angoumois grain moth (only the Angoumois grain moth is an internal feeder).

Secondary insect pests: Secondary pests generally feed on grain that is going out of condition or damaged. Damaged grain kernels have exposed endosperm that is accessible food for stored grain pest and fungi. It includes drugstore beetle, spider beetles, white marked spider beetle, hairy spider beetle, cigarette beetle, silken fungus beetles atomaria, silken fungus beetles cryptophagus, dermestids, black carpet beetle, larder beetle, glabrous cabinet beetle, mottled dermestid beetle, ornate carpet beetle, , plaster beetle, spotted hairy fungus beetle, hairy fungus beetle, sap beetle, foreign grain beetle, warehouse beetle, lesser mealworm, broadhorned flour beetle, dark mealworm, yellow mealworm, black flour beetle, american black flour beetle.

Stored-product moths: White shouldered-house moth, brown house moth, almond moth, mediterranean flour moth, indian meal moth, meal moth, european grain moth, clothes moths.

Other insects: Psocids, silverfish or firebrat.

Non insect pests: Rodents, grain mite and birds.

Major insect pests of cereal grains: Larger grain borer (*Prostephanus truncates*), lesser grain borer (*Rhizopertha dominica*), rice moth (*Corcyra cephalonica*), rice weevil (*Sitophilus oryzae*), granary weevil (*Sitophilus granaries*), maize weevil (*Sitophilus zeamais*), khapra beetle (*Trogoderma granarium*), flat grain beetle *(Cryptolestus minutas),* angoumois grain moth (*Sitotroga cerealella*).

Major insect pests of pulses/ grain legumes: Pea pulse beetle (*Bruchus pisorum*), Lentil pulse beetle (*Bruchus lentis*), gram, mung, urd, lobia, moth, pea, arhar, masur and pea pulse beetle (*Callosobruchus maculates*), Hourse gram pulse beetle / gram dhora (*Callosobruchus chinensis*), lobia and mung pulse beetle (*Callosobruchus analis*), hyacinth bean and arhar pulse beetle (*Callosobruchus theobeamae*) arhar pulse beetle (*Callosobruchus phaseoli*), rajmas pulse beetle *(Zabrotes subfaciaatus)* and groundnut and imli pulse beetle (*Caryedon serratus*).

Insect pests of mill and milled products: Flour mill beetle (*Cryptolestus turcicus*), merchant grain beetle (*Oryzaephilus mercator*), saw toothed grain beetle (*Oryzaephillis surinamensis*), longheaded flour beetle (*Latheticus oryzae*), red flour beetle (*Tribolium castaneum*), black flour beetle, American black flour beetle (*Tribolium audax*), confused flour beetle (*Tribolium confusum)* and Indian meal moth (*Plodia interpunctella*), brown house moth (*Hofmannophila pseudospretella),* almond moth (*Sitotroga cerealella)*.

Insect pests of Spices and condiments: Cigarette beetle (*Lasioderma sericorne*), Drugstore beetle (*Stegobium paniceum*), Silken fungus beetles (*Cryptophagous* spp).

Insect pests of dry fruits: European grain moth (*Nemapogon granella*), Almond moth (*Ephestia cautella*), Mediterranean flour moth (*Ephestia kuehniella*) and White-shouldered house moth (*Endrosis sarcitrella*).

Integrated Pest Management (IPM): Integrated Pest Management involves understanding interactions between stored product environment and insects associated with stored products, and replacing all or most of the chemical applications with cost-effective non-chemical alternatives. IPM in storage has been classified in two ways i.e. (1) preventive and (2) curative. The Cracks and crevices are probable sites for "Cross infestation", the containers and bags with eggs and larvae hidden in the mesh of bags and the trucks, trolleys and bullock carts used for transportation of grains.

Preventive

(A) Sanitation/Cleaniness: Clean the threshing floor/yard and machines like harvester, thresher, trucks, trolleys or bullock carts free from infestation before their use.

(B) Removal of all the old loose and mud plaster. Keep bagged grains/seeds at distance from walls for inspection and fumigation and avoid the grains/seeds to absorb moisture from moist surfaces. Clean the stores and receptacle by keeping bags on dunnage.

(C) Apply new mud plaster to fill all the cracks and cravices before filling the grains in storage structures / godowns.

(D) Legal method: Imposition of Destructive Insect Pests Act 1914 for prevention of entry of an insect not found in a particular area.

(E) Disinfect the stores and receptacles by spraying the 0.50% Malathion 50% EC or 0.50% Primiphos methyl or 0.5% DDVP or 0.02% Pyrethrin @ 3L/100 m^3.

(F) Use new gunny bags for fresh harvest/grains or treat gunny bags by dipping of 10-15 minutes in any one of the following insecticidal 0.1% Malathion or 0.01% Fenvelrate 20 EC or 0.01% Cypermethrin.

(G) Maintain the moisture below 10 % in grains before storage.

(H) Use 7 tablets (3g each) of Celphos/Phosfume (aluminium phosphide) per 1000 cubic feet of empty space with 7 days of exposure period.

Curative

Non-chemical control measures: (1) Ecological control measures (2) Physical control measures (3) Cultural control measures (4) Mechanical control measures (5) Use of botanicals (6) Engineering control measures (7) Chemical control measures (8) Traps

1 Ecological control measures

The pest population in store can be checked through temperature and grain moisture regulation.

Low temperature: Producers and grain handlers can use ambient cold air to prevent or control infestations. Stored product insect pests generally do not feed or reproduce at temperatures below 18°C. Lower temperatures can also be used to cause mortality. For example, grain kept at -5°C for 12 weeks will control stored insect pests at all the life stages. Mixing and transferring infested grain from one bin or pile to another may reduce the temperature of stored grains. Aeration systems are very effective to reduce grain temperature as well as reducing moisture migration. A Constant grain temperature of -5°C for 12 weeks, -10°C for 8 weeks, -15°C for 4 weeks, and 20°C for 1 week is required to control stored grain insect pest infestations. Immature stages of almost all insect pests die below 14^{0}C. Death occurs rapidly at freezing point -10^{0}C to -20^{0}C. Aeration cool ambient air results in low temperature. Bulk grains can also be cooled by refrigeration in cold stores (Banks and Field, 1995). Some species are able to tolerate cold and some are resistant. In general *Tribolium casteneum*, *T. confusum* and *Oryzaephilus mercator* most cold susceptible and *T. granium*, *S. granarius*, *E. cutella* and *P. interpunctella* are most cold resistant (Field,1992).

High temperature: High temperature from 50°C to 60°C for 10-20 minutes lethal to all insects. Pupal stage of *Rhizopertha dominica* is heat tolerant. Muthu and Majumdar (1974) found that most of the insects attacking stored seed spices die when the body temperature reaches upto 60°C but in case of infrared, when the temperature attained is 65-68°C in the rotating aluminium or stainless tubes give 100 per cent mortality of drug store beetle, *Stegobium paniceum*. After applying heat treatment, grain must be immediately cooled so that it does not overheat that may cause a new insect infestation. Energy costs for this method of control are high. Therefore, use of grain dryer is recommended only when you need to dry the grains rapidly.

Solar Bed: Three times exposures on solar bed during sunny days for 15 minutes before storage (first exposure) and at regular intervals of 60 days (second exposure) 120 days (third exposure) during entire storage period of 180 days respectively is found effective.

Means of heating: Hot air fluidized bed, Infra red, High Frequency dielectric and microwave heating

Grain moisture: Grain moisture regulates biological activity of insects in storage. Atmospheric humidity and grain moisture are closely related factors. Even though stored grains are dried to safe moisture level before storage, moisture content in due course comes to equilibrium with humidity in the air. Paddy at 13 per cent moisture does not allow insects to multiply in storage (Hall, 1980). Grains stored at around 10 per cent moisture content escape from the attack of insects (except 'khapra' beetle).

2 Physical control measures

Controlled atmosphere (CA) disinfestations technology involves the alteration of natural storage gases, CO_2, O_2 and N_2 present in a storage space so as to obtain an artificial atmosphere that prevents multiplication of insects, mould growth and quality deterioration of food grain. Maintaining O_2 level below 1% and CO_2 will automatically high which will be lethal to all stages of insects referred as "Hermetic Storage" (Benks and Annis, 1990).

3 Cultural control measures

Pulse beetle attacks whole pulses only. Split pulses escape the attack of *Callosobruchus* spp.

Store the food grains in air tight sealed structures to prevent the infestation by insects.

4 Mechanical control measures

Screening of grains: Broken and cracked grains promote the attack of stored grain pests. Hence, screening/sieving out of such grains to eliminate the condition which favour storage pest infestation. Screening should be done regularly. Bags used for carrying the screenings should not be used again unless disinfected.

Monitoring stored grain for insect pest infestations: To prevent insect infestations is to monitor bin-stored grain every two weeks to detect early signs of deterioration or infestation.

5 Use of botanicals

Botanicals are naturally occurring chemicals having insecticidal properties derived from plants sources.

Plants and other non-toxic grain protectants / Use of Edible oils: Biopesticides such as *Neem Azadirachta indica* (Leaf, seed, oil, kernel powder, and crude extract) possess repellant, antifeedant and feeding deterrent properties against storage insects (Savitri and Rao, 1976). *Neem* seed oil 1%, *neem* seed kernel powder 4%, and *mahua* oil 1% proved repulsive, potent, oviposition inhibitor in checking damage by *Tribolium castaneum* upto 8 months (Singal and Chauhan, 1997).

Mustard and groundnut oils @ 7.5 ml/kg of stored grains are effective for protection of stored pulse grains against pulse beetle 'dhora' upto 9 months. Pulse grain protectants namely groundnut oil @ 3.75 ml/kg+ turmeric powder @ 1.75 g/kg, mustard oil @ 3.75 ml + turmeric powder @ 1.75 g/kg, neem oil @ 10 ml/kg, 7cm sand and dung cake ash covering on gram proved effective in protecting the treated gram grain from the infestation of pulse beetle, *Callosobruchus* spp. for a period of 180 days. The use of coconut, mustard, soyabean, groundnut and rapeseed oil is found effective @ 5ml/kg for checking pest infestation upto six month of storage (Singal, 1990).

6 Engineering control measures

Use of improved storage structures: The traditional receptacles for food grains storage are not resistant to insects, rodents, fire hazards, moisture and are also not suitable for fumigation/ disinfestations during storage.

Improved traditional storage structures: Use of improved structures i.e. pusa bin, gade, puri, kothi, patara, pucca kothi and Gharelu theka.

Modern storage structures: Food storage receptacles constructed on scientific basis and techniques can store food grains in safe and sound conditions for long periods. Such scientific structures are flat and hopper bottom-metal bins, Composite bins, Partly underground and above ground structures, Seed storage bins, Ferro-cement bins, Pusa bin, Improved godowns, Bulk storage installations and Vacuum process storage.

Urban bins (Circular or Square in shape): Gunny bags with polyethylene lining or polyethylene bags of 300-400 gauge thickness also protect paddy from cross infestation of insects upto 2 years as well as stop moisture migration to the paddy stored in these bags. However, *R. dominica* and *S. oryzae* can penetrate polyethylene lining. Sometimes the loss of seed viability was also greater in polyethylene lined bags than in jute bags without polyethylene (Singal, 2006).

7 Chemical control measures

Inspite of regular prophylactic measures infestation develops due to hidden infestation etc. so some curative measures have to be adopted with chemical applications.

Knockdown chemical: Use knockdown chemicals like pyrethrum spray, lindane smoke generator or fumigant strips useful against flying insects as well as hidden insects in cracks and cravices. Treat the walls, dunnage materials and ceilings of empty godown with malathion 50 EC @10 ml/L.

Grain protectants: The grains meant for seed can be protected by mixing of malathion 5% @ 250g/q or deltamethrin 2.8 EC @ 4ml/q seed.

Fumigants: Fumigation with aluminium phosphide (Phostoxin/ Celphos) @ 1 tablet (3gm each) for 1 tonne (10 bags) or 7-10 tablet (3gm each) 1000 cubic feet space with exposure period of at least 7 days.

Ethylene dibromide (EDB) ampule is very toxic to all the stages of insects. The doses 3ml/q for wheat and pulses and 5ml/q for rice and paddy. EDB also applied @1.7 litres for 1000 cubic feet space with exposure period of 7 days.

Ethylene dichloride carbon tetrachloride (EDCT) is used as mixture in the ratio of 3:1 (v/v) for large scale fumigation @ 35 litres/100 cubic meter space in large scale storage with exposure period of 4 days. Use EDCT mixture @ 55 ml/q or one liter for 20q in small storage food grains irrespective of bulk or bag storage.

Note: Do not mix any insecticide dust with grains. The fumigants should be used only in airtight stores by specially trained personnel. Avoid use of EDB in grains meant for seed purposes.

8 Traps

Use of pit fall traps, probe traps, sticky traps, light traps and artificial crevices are found very effective as they indicate the early presence of insect moths, early detection of infestation and improve the effectiveness of insecticidal application in storage structures.

References

Benks, H.J. and Annis, P.C.1990. Comparative advantage of high CO_2 and low O_2 types of controlled atmospheres for grain storage, pp. 93-122. In: *Food Preservation by Modified Atmospheres,* M. Calderon and R. Barkai Golan (eds.), CRC Press, Roca Baton, FL.

Banks, H.J. and Fields, P.G, 1995. Physical methods for insect control in stored-grain ecosystems. In: Jayas DS, White NDG, Muir WE, eds. Stored-grain ecosystems. New York: Marcel Dekker, 353-409.

Field, P.G. 1992. The control of stored product insects and mites with extremes temperatures. *J. Stored. Prod. Res.* 28: 89-118.

Hall, D.W. 1970. Handling and storage of food grains in tropical and subtropical areas. FAO. Agricultural Development Paper No. 90 and FAO, Plant Production and Protection Series No. 19, pp-350, FAQ, Rome.

Khare, B.P.1994. Stored Grain Pests and Their Management. Kalyani Publishers.

Muthu, M. and Majumdar, S. K. 1974. Insect control in spices. Proceedings of a symposium on Development, prospects for spice industry in India, held by Assoc. Food Sci. and Technology. In Mysore, 1974, 35.

Pruthi, H. S and Singh, M. 1950. Pests of stored grain and their management. *Indian Journal of Agricultural Science* 18 (4) : 1-86. (Special issue).

Pruthi, H.S. and Singh, M. 1948. Pests of stored grain and their control. *Indian Council of Agricultural Research*, New Delhi. India. 88pp.

Savitri, P. and Rao, S.C. 1976. Studies on the admixture of neem seed kernel powder with paddy on the control of important storage pests of paddy. *Andhra Agric. J.* 23: 137-143.

Singal, S.K. and Chauhan, R. 1997. Effect of some plant products and other materials on development of pulse beetle, *Callosobruchus chinensis* (L.) on stored pigeon pea, *Cajanus cajan* (L.) Millsp. *Journal of Insect Science*. 10, 196–197.

Singal, S.K. 2006. Stored grain pests and their management. Emerging trends in economic entomology, etd by, B.S. Chhillar et.al. Centre of advanced studies, Department of Entomology, CCSHAU, Hisar.

Singal, S.K. and Singh, Z. 1990. Studies of plant oils as surface protectants against pulse beetle, *Callosobruchus chinensis* (L.) in chickpea *Cicer arietinum* L. in India. *Tropical Pest Management*, 36(3): 314-316.